A NOTE ON THE AUTHOR

Kathryn Harkup is a chemist and author. Kathryn completed a doctorate on her favourite chemicals, phosphines, and went on to further postdoctoral research before realising that talking, writing and demonstrating science appealed more than hours slaving over a hot fume-hood. Kathryn is now a science communicator, giving regular public talks on the disgusting and dangerous sides of science. Kathryn has written several books; her first was the international bestseller *A is for Arsenic*, which was shortlisted for both the International Macavity Award and the BMA Book Award.

@RotwangsRobot

Also available in the Bloomsbury Sigma series:

Sex on Earth by Jules Howard
Spirals in Time by Helen Scales
A is for Arsenic by Kathryn Harkup
Suspicious Minds by Rob Brotherton
Herding Hemingway's Cats by Kat Arney
The Tyrannosaur Chronicles by David Hone
Soccermatics by David Sumpter
Goldilocks and the Water Bears by Louisa Preston
Science and the City by Laurie Winkless
Built on Bones by Brenna Hassett
The Planet Factory by Elizabeth Tasker
Catching Stardust by Natalie Starkey
Nodding Off by Alice Gregory
The Edge of Memory by Patrick Nunn
Turned On by Kate Devlin
Borrowed Time by Sue Armstrong
The Vinyl Frontier by Jonathan Scott
Clearing the Air by Tim Smedley
Superheavy by Kit Chapman
The Contact Paradox by Keith Cooper
Life Changing by Helen Pilcher
Our Only Home by His Holiness The Dalai Lama
First Light by Emma Chapman
Ouch! by Margee Kerr & Linda Rodriguez McRobbie
Models of the Mind by Grace Lindsay
The Brilliant Abyss by Helen Scales
Overloaded by Ginny Smith
Handmade by Anna Ploszajski
Beasts Before Us by Elsa Panciroli
Our Biggest Experiment by Alice Bell
Worlds in Shadow by Patrick Nunn
Aesop's Animals by Jo Wimpenny
Fire and Ice by Natalie Starkey
Sticky by Laurie Winkless

DEATH BY SHAKESPEARE

Snakebites, Stabbings and Broken Hearts

Kathryn Harkup

B L O O M S B U R Y S I G M A
LONDON • OXFORD • NEW YORK • NEW DELHI • SYDNEY

BLOOMSBURY SIGMA
Bloomsbury Publishing Plc
50 Bedford Square, London, WC1B 3DP, UK
29 Earlsfort Terrace, Dublin 2, Ireland

BLOOMSBURY, BLOOMSBURY SIGMA and the Bloomsbury Sigma logo are trademarks of Bloomsbury Publishing Plc

First published in the United Kingdom in 2020
This edition published 2022

A catalogue record for this book is available from the British Library

Library of Congress Cataloguing-in-Publication data has been applied for

ISBN: PB: 978-1-4729-5820-4; eBook: 978-1-4729-5824-2

2 4 6 8 10 9 7 5 3 1

Chapter illustrations by Ele Willoughby

Typeset by Deanta Global Publishing Services, Chennai, India
Printed and bound in Great Britain by CPI Group (UK) Ltd, Croydon CR0 4YY

Bloomsbury Sigma, Book Fifty-one

To find out more about our authors and books visit www.bloomsbury.com and sign up for our newsletters

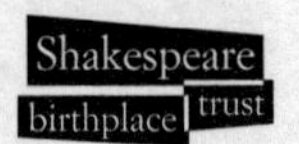

In partnership with the Shakespeare Birthplace Trust

To my Gran

Contents

> I shall offend, either to detain or give it. The contents, as in part I understand them, are to blame.
>
> *King Lear*, Act 1, Scene 2

Prologue 9

Chapter 1: Our Humble Author 11

Chapter 2: All the World's a Stage 43

Chapter 3: Will You Be Cured of Your Infirmity? 71

Chapter 4: Off With His Head! 103

Chapter 5: Murder, Murder! 139

Chapter 6: The Dogs of War 171

Chapter 7: A Plague O'both Your Houses! 205

Chapter 8: Most Delicious Poison 237

Chapter 9: To Be, or Not to Be 269

Chapter 10: Excessive Grief the Enemy to the Living 293

Chapter 11: Exit Pursued by a Bear 313

Epilogue 337

Appendix 341

Bibliography 353

Acknowledgements 361

Index 363

Prologue

All the world's a stage

As You Like It, Act 2, Scene 7

William Shakespeare occupies a special place in history. No other writer has enjoyed the same success and prolonged reverence as the Bard of Avon. He has entertained us for over four centuries, given us a wealth of cultural references and inspired countless adaptations and interpretations of his work.

He may have been catering to the tastes of his audiences in sixteenth- and seventeenth-century London but his plays and poems are still known and enjoyed today, and not just in England but around the globe. Shakespeare tapped into something that crosses cultural and linguistic borders. The setting and the century – ancient Egypt, a medieval battlefield in France or an enchanted island somewhere in Renaissance Europe – may not be easy to relate to, but his themes of love, hatred, jealousy and bereavement certainly are.

The Bard drew from a vast world of history, literature, imagination and everyday experience to produce plays that make us laugh, cry, gasp and think. His ability to take simple ideas and weave into them such complexity of detail and depth of character made his plays stand out among those of his contemporaries. His eye for detail has convinced many that he must have been a scholar of such varied fields as law and medicine, as well as spending time travelling on ships and soaking up the atmosphere of Italy.

We have no proof that he did any of these things. His ability to absorb information, thread it into plays and present it in such a beautiful way has perhaps fooled us into thinking he had a more detailed knowledge and varied experience of life than he really did. However, two subjects he was undoubtedly an expert in, having experienced both at close quarters, was life and death in Elizabethan and Jacobean England.

Shakespeare understood death in a way that perhaps we don't today. The playwright lived in a time when lives were often short and death was a social event. He may have understood little about the science of the process of death but he knew what it looked, sounded and smelled like. Today death is sanitised, screened off and seldom talked about. Often the detail is hidden from us completely. People living in the sixteenth and seventeenth centuries visited the sick and dying and were personally involved in caring for friends and relatives in their last moments. They also witnessed public executions, saw street brawls and lived in constant fear of visitations from the plague.

With limited effective medical treatments available, the grim reality of death, from even the most trivial of illnesses and infections, was well known, up close and in detail. Death was a familiar feature of everyday life – 'To die is as common as to live' (*Edward III*) – and Shakespeare didn't shy away from describing it. Death was simply part of the richness of life that he wrote about so brilliantly. Spectacular deaths, noble deaths, tragic deaths and even mundane deaths are all included in his plays, sometimes in astonishing detail. This book will explore them all.

CHAPTER ONE

Our Humble Author

One man in his time plays many parts

As You Like It, Act 2, Scene 7

Shakespeare has given us a wealth of written material to enjoy, but there are few traces of the man himself. His life began as it continued, in historical vagueness. Even his date of birth is a matter of some dispute. William Shakespeare was born on either 21, 22 or 23 April, but his birthday is generally celebrated on 23 April, to coincide with St George's Day, England's patron saint, as well as the day of his death 52 years later. We do know that his life began in Stratford-upon-Avon and that he was baptised on 26 April 1564.

William was one of eight children born to John and Mary Shakespeare. Of their eight children only five survived to adulthood (in order of appearance, Joan, William, Gilbert, Richard and Edmund). William outlived all his brothers, a remarkable feat given when he was born. From the very start of his life death was never far away.

On 11 July 1564, just under three months after the playwright's birth, the parish register at Holy Trinity Church, Stratford, contains the ominous phrase '*Hic incepit pestis*' (here begins plague). This particular outbreak claimed one-sixth of Stratford's population, at least 200 people, 10 times the normal mortality rate. The dead included nearly two-thirds of all children born in the town that year. Shakespeare was one of the lucky ones. Perhaps early exposure to the disease gave him some protection from future infections in later life and helped him survive the regular plague outbreaks in London in the sixteenth and early seventeenth centuries. During this 1564 outbreak, children and the elderly were particularly vulnerable, but everyone was terrified. In August the town council held an emergency session to discuss organising relief for the victims, but they sat outside in the fresh air of the chapel garden to try to limit their chances of catching the disease.

The population of the whole of England in the late sixteenth century was only around four million.* And, against the odds, the population managed to increase by roughly 1 per cent each year of Elizabeth's reign. This was thanks to an exceptionally high birth rate, which was needed to keep up with and exceed the considerable death rate.

Average life expectancy in the Tudor period was just 38, and the most dangerous time of life was the first few years. Just being born was a risky business for both mother and child. Midwifery was the sole province of women at the time, and though many would have been

---

* All population numbers are approximate and offered as educated guesses. There was no official census in Renaissance England and birth and death records are not complete.

very experienced, there was no specific training. Although apprenticing was common, few bothered to register with the Church Courts to obtain a midwife's licence, even though it required no examination or test of knowledge. A midwife's experience would have been invaluable during straightforward births, but there was little that could be done for anything even slightly out of the ordinary. Breach births, difficulties in pushing the child through the birth canal,* and haemorrhage, as well as the risk of postpartum infection, were all complications that could prove fatal.†

If the child made it out into the world alive there was a 9 per cent chance it would die in the first week of life and a further 11 per cent chance that it would die before the end of the first month. A less than 80 per cent chance of living to see your first birthday is poor by any standards, especially compared to today's figure of just under 97 per cent of babies surviving to the age of one worldwide (though there is significant variation around the globe). But the danger wasn't over: a third of children died before their tenth birthday. Hence surviving to adulthood was something of an achievement. Genetic factors would have improved some infants' chances, but being born to a prosperous family certainly helped their odds of survival. Shakespeare was probably lucky on both counts.

Despite the appalling child mortality rates, England was a country of young people, with half of the

---

* Forceps were not in use in England until the mid-eighteenth century.

† The mothers of several Shakespearean characters die in this way, perhaps most significantly in *A Midsummer Night's Dream* where Titania and Oberon fight over the custody of a changeling boy whose mother died in childbirth.

population under 25. One diarist considered that 40 was the year which 'begins the first part of the old man's age'. The longer you lived, the better your chances of surviving even longer. Those that made it to 30 would likely see 60. Shakespeare lived to see just the first day of his 53rd year, a very respectable age for the time. However, we know relatively little about what he did for many of those 52 years.*

Shakespeare's early life was probably spent in Stratford, first attending the local grammar school and later most likely working in his father's business (his father was probably a glover). One fact is certain: at the age of 18 William married the 26-year-old Anne Hathaway. The wedding seems to have been rather hastily arranged, probably because Anne was pregnant at the time.

Together William and Anne had three children, Susanna, born in 1583, and the twins, Judith and Hamnet, born in 1585. Hamnet died aged only 11, of causes unknown, but Shakespeare's two daughters both lived long lives, Susanna until she was 66, and Judith until she was 77. All were born in Stratford and William was presumably present at their baptism in the local parish church. But what Shakespeare was doing for the seven years after the birth of the twins is a complete mystery. Although there are many theories and much speculation over the missing years, the next we know for certain is

---

* Some have taken this lack of information as evidence that he was not the author of the works usually attributed to him. However, more is known about Shakespeare's life than any of his contemporary playwrights (with the possible exception of Ben Jonson, who did much to preserve his legacy).

that by 1592 Shakespeare had established himself in London as a playwright.

★ ★ ★

Shakespeare's taste for the theatrical life may have started young. Staged productions were a popular form of entertainment in England long before Shakespeare entered the scene. Amateur entertainments would be put on during festivals, and larger towns like York and Lincoln performed extended productions, or a series of linked performances, called mystery plays. These involved large numbers of the local populace and acted out episodes from the Bible. Perhaps Shakespeare got a taste for the stage while performing in one of these events in his youth. However, by the time he was born, performances of mystery plays were declining.*

There were also professional acting troupes touring the country giving performances wherever and whenever they could. These companies arrived in town and set up their stage in any available location, often, as depicted in *Hamlet*, arriving at the house or castle of a local nobleman. It was hard work. Everything, from costumes and props to musical instruments, had to be carried with them. Venues varied considerably and the text and staging had to be adapted accordingly. There would have been little time to rehearse and the income was anything but steady.

---

* These epic religious dramas were suppressed by Queen Elizabeth's Protestant authorities as they seemed more in keeping with Catholic practices.

After a free performance in front of the mayor or other local official (to get his authorisation), further plays were acted out for paying audiences. The money, however, was not always forthcoming. At Norwich, in 1582, one man refused to pay for his ticket until he had seen the play. A fight broke out involving several of the actors as well as members of the audience and a man bled to death after he was struck with a sword. Life, and death, on the road could be tough. Nevertheless, acting companies continued to tour throughout Shakespeare's lifetime and beyond, especially in the summers and when they were forced out of the capital because of plague. But to make your name in the theatrical world you had to go to London.

How Shakespeare made the transition from life in a bustling market town to fame and fortune in the capital city is something of a mystery. One theory has it that the Queen's Men,* when they visited Stratford in 1587, were missing an actor and the 20-something William Shakespeare joined their ranks. The company had stopped in Thame in Oxfordshire when a fight broke out between two of their members, William Knell and John Towne. Knell was stabbed in the neck and died of his injury, leaving a gap in the company that some think was filled by Shakespeare.

The problem with this theory is that no one knows the route the company took on their tour and whether they stopped at Stratford before or after Knell was fatally stabbed. There are also no known professional links between Shakespeare and the Queen's Men and it was

* Acting companies came under the patronage of wealthy aristocrats and each troupe was known by their patron's name.

unlikely they would have recruited actors while on the road.

★ ★ ★

London was undoubtedly the place to be if you wanted a career as an actor or playwright, but it had its risks. England's capital had a well-known and well-deserved reputation for being an unhealthy place to live. Life expectancy in England may have averaged 38 but it varied greatly across the country, and people living in the countryside had a much better chance of experiencing old age than those in the city. Average life expectancy in the best London districts was 35 and in some of the poorer areas it struggled to reach above 25. Threats to Elizabethan life were many and varied. To survive, you had to be both careful and lucky.

The death rate in the capital consistently outstripped the birth rate, but the population continued to grow as people swarmed in from the countryside and refugees from the continent disembarked at the docks. In the 1590s it was one of the largest cities in Europe – only Paris and Naples were bigger. Its streets teemed with life. At the beginning of the sixteenth century there were approximately 50,000 people living in the capital, but a century later that number had quadrupled to around 200,000.

Greater London, including Westminster and Southwark, stretched over only a few square miles, and the physical growth of the central part of the city was severely restricted, so every available scrap of land was used for building. Streets got narrower; light and fresh air disappeared as homes were squeezed closer and closer together. Churches profited by renting out their churchyards for building small tenements. Many parishes then had to bury their dead in the churchyard of St Paul's Cathedral. Coffins were stacked on top of each other and squeezed into every available spot. A spade thrust into the ground to create a new plot often cracked open a coffin lid lying just below the surface. St Paul's churchyard also doubled as a marketplace and was the centre for a burgeoning publishing industry. Everywhere was crowded.

The fabric of the city itself was one potential threat to life. Construction was unregulated, there were no minimum building standards to adhere to or prescribed safety regulations to comply with. Buildings were mostly made of wood and were heated with open fires.

Outside the buildings rubbish was dumped in the narrow streets. The occupants of a building were supposed to keep their frontage clean and presentable, on pain of a stiff fine, but many ignored this instruction, as well as the fine. William Shakespeare's father John was one of those found guilty of having an 'unauthorised midden heap' in Henley Street, Stratford. He should have taken his waste to the communal heap further away, but he did at least pay the shilling fine.

Piles of domestic, animal and human waste accumulated in the streets. Blood and offal from the slaughterhouses clogged up the gutters and fed ravens and stray cats and dogs. The rubbish also attracted rats, and with them, disease. The hope was that rain would wash the detritus into ditches and rivers but, even though England has something of a reputation for wet weather, the downpours were rarely sufficiently torrential to keep the streets clean.

In London the Fleet Ditch and Moor Ditch funnelled the foul waste into the Thames, but periodically they became blocked. So bad was the resulting stench that 'idlers' and vagrants were set to work to clear them out. It can be no surprise that dead dogs floated in the Thames along with all the other waste that had washed into it from the open sewers.

It was from the Thames that many Londoners obtained their drinking water and fish from the river were sold and eaten. Where once there had been fish in the Town Ditch that ran around the City's walls, there were now only bacteria and other parasites swimming in its filthy waters. Typhoid outbreaks would have been a regular occurrence, although typhoid was not then recognised as a disease. It rarely observed the social niceties of the age and killed rich and poor alike. It was typhoid that caused the death of young Prince Henry, King James's son and the popular heir to the throne, in 1612.

It wasn't just eating the local fish and drinking the local water that could make you seriously ill. There were no food regulations. Unless food was obviously rotting, it could be sold without consequence to the seller but with considerable risk of food poisoning to the consumer.

London was also a major port where ships embarked and goods and people from all over the world disembarked. The opportunities to introduce new diseases and spread new infections to the local population were plentiful. The source of infections was not well understood and disease was commonly thought to be caused by bad odours. Considering the stink from the streets and the generally poor health of the population, it was an easy connection to make, even though it was incorrect. From this association came the Elizabethan idea of sanitary conditions being linked to sweet-smelling air. To prevent infection, perfumes and pomades were used to anoint the body and pomanders made from herbs, or oranges stuck with cloves, were held under the nose. Rooms, houses and sometimes streets were fumigated by boiling vinegar or burning pitch. Naturally, it didn't help.

To make matters worse, good personal hygiene was decidedly out of fashion. Clothing was seldom changed. Bathing was infrequent and usually only undertaken as a cure for gout, rheumatism or 'to amend your cold legs against the winter'. Instead of bathing, people rubbed themselves down with a coarse cloth. Hands and faces received special attention and were washed at least daily. There were abundant opportunities to die from contagious disease in Elizabethan England. It can be no surprise that all of Shakespeare's plays contain references to some disease or other.

* * *

The 20-something Shakespeare arrived in this unwholesome environment and started to make himself known on the theatrical scene. There were a number of notable playwrights in London when Shakespeare got there, including Christopher Marlowe, Robert Greene and Thomas Nashe. It would have been difficult for him to stand out in such a crowded and prestigious field, but he evidently did. The success of the newcomer caused some consternation among the established theatre professionals, if the cutting comments made by Robert Greene are anything to go by. It is from Greene that we have the wonderful phrase 'upstart crow', assumed to be a snipe at

Shakespeare although he isn't specifically named in the text. The full venom-filled quote runs thus:

> Yes, trust them not, for there is an upstart Crow, beautified with our feathers, that with his Tigers heart wrapt in a Player's hide, supposes he is well able to bombast out a blank verse as the best of you: and being an absolute *Johannes factotum* [literally John do-all, or Jack-of-all-trades], is in his own conceit the only Shake-scene in a country.

From Greene's point of view Shakespeare had many faults. He had taken material from other playwrights (though plagiarism wasn't really a concept at the time and borrowing ideas was common and largely accepted); he was an actor *and* a playwright; and he came from the countryside. Like most of his fellow playwrights, except Shakespeare, Greene was university educated. The Bard's alleged poor education has been seen as suspicious by those who did not believe the grammar-school boy from Stratford could have written the plays normally attributed to him. His supposedly poor learning also drew contempt from other contemporary playwrights – Ben Jonson derided Shakespeare for his 'small Latin and less Greek'. But a sixteenth-century grammar-school education provided a better knowledge of both languages than might be expected of modern university graduates in these subjects. Grammar-school curricula of the time focused on studying Greek and Roman plays, oratory and rhetoric. It would have been excellent training for a playwright.

Greene was a noted writer, mostly of romantic novels, but also of plays. His output was considerable, and needed to be to fund his profligate lifestyle. In keeping with many of his contemporaries he seems to have lived life to the full and enjoyed the seamier side of London's many attractions. He abandoned his wife and took up with a mistress named Em, who was the sister of a notorious thief and cut-throat named 'Cutting Ball' who was hanged at Tyburn for his crimes.

Greene earned little and spent extravagantly on clothes and entertainment. It was probably his excessive lifestyle that brought about his downfall, one way or another. The rumour at the time was that he had contracted his fatal illness as a result of an excess of pickled herring and Rhenish wine that he had indulged in when out with his fellow cronies. Although he lingered for another month after his spectacular feast, he was abandoned by his friends and attended only by flocks of lice and a landlady who took pity on him. He finally succumbed on 3 September 1592 to what was probably tertiary syphilis, aged only 34.

However, there are alternative theories for Greene's demise. One comes from a medical report supplied by Cuthbert Burby, which states that Greene complained of pains in his belly. The swelling observed in his abdomen and face suggested dropsy (oedema) caused by cirrhosis of the liver; the Rhenish wine may have been a contributing factor after all in a case of chronic alcoholism. Whatever the cause, Greene died miserable, poor and alone and it is unsurprising that he spat out his scathing attack on another playwright on his deathbed.

Of all the dread diseases that could have killed Greene and others living in London's unhealthy streets, most feared of all was the plague. Yet another theory has it that Greene was an early victim of the plague epidemic that broke out just days after his death. The outbreak of 1592–3 began just as Shakespeare was making his name as a man of the theatre, and could have ended his career before it really got started. It closed the theatres and put the theatrical industry under considerable financial strain.

★ ★ ★

London was rarely plague-free but there were several severe epidemics during Shakespeare's time there. Plague outbreaks not only threatened the lives of Londoners but also seriously curtailed the incomes of actors and playwrights. Theatres were closed as a precaution against the spread of the disease because they attracted huge crowds of people in close proximity. In 1574 the authorities forbade playing 'in the time of common prayer, or in the time of great and common plague'. By 1584 it was specified that playhouses should not open until the number of plague victims had fallen below 50 per week for three weeks. In 1604 the Privy Council reduced this number to 30. Although the limits were not always strictly observed, it nonetheless had a crippling effect on the theatrical profession.

The 1592–3 epidemic claimed the lives of at least 10,000 Londoners and probably many more (according to the Bills of Mortality the number could have been as high as 18,000). All the playhouses were closed for over a year, with only a few brief re-openings. During this time acting companies struggled to survive. When the theatres were closed, acting troupes would head out on tour to try to eke out a living until the pestilence subsided. The epidemic abated enough to open the playhouse doors again for the Christmas season of 1593–4, but it wasn't until the autumn of 1594 that the normal rhythm of London life resumed.

Necessity rather than choice perhaps forced Shakespeare into a new venture as a poet. He was lucky to find tremendous success, both artistically and financially, in this alternative career to support what must have been a meagre income from touring. This is when Shakespeare produced the narrative poem *Venus and Adonis*, complete with a gushing dedication to Henry Wriothesley, the third Earl of Southampton. Had Shakespeare succumbed to the plague at this time he would have been known to posterity as a minor playwright, who produced one notable poem. His theatrical work would have been completely overshadowed by his contemporaries Marlowe, Jonson and Greene. But he survived. Shakespeare's work has infiltrated our language and culture to such an extent that it is difficult to imagine what the modern world would be like without his influence.

Plague was the cause of the considerable reshuffling of actors between companies. Actors died, or left the profession to seek a more reliable income. In some cases the wealthy patrons themselves died, leaving companies in need of a new sponsor and a new name.

Pestilence was not the only threat to the success of an acting troupe. The Earl of Derby's Men were forced to disband when their patron, Ferdinando, Lord Strange, by then the Earl of Derby, died in 1594 under mysterious circumstances. The previous year Strange had received a letter from Richard Hesketh asking him to stand as leader of a plot against the Queen. Rather than take up the proposal, Strange denounced Hesketh to the authorities. But that wasn't the end of the story. The following year, when Strange died suddenly, rumours of witchcraft, or perhaps poisoning, circulated. Now the acting troupe was left without a sponsor. The remaining members of the troupe were eventually amalgamated into the Lord Admiral's Men. The pressures on an acting company were great and, for some, too much. Many acting troupes simply disbanded at this time.

The 1592–3 plague outbreak so disrupted the playing companies that when it was all over, and the playhouses in London reopened, only two major companies were left. One troupe, the Lord Admiral's Men, survived the plague, but not without the group's membership undergoing some major reshuffling. They were led by the actor, property owner, brothel keeper and impresario, Edward Alleyn.

A second company was formed by Henry Carey, the Lord Chamberlain, and was therefore known as the Lord Chamberlain's Men. The Lord Chamberlain was head of the royal household and so Carey was in charge of entertainment for the Queen (among other duties). His company of actors became the preferred troupe called upon to entertain the monarch. They were led by actor, theatre owner and great rival to Edward Alleyn, Richard Burbage. This was Shakespeare's company, the only troupe with whom he would ever act and the only one for whom he would write plays.

Few other playwrights wrote exclusively for one company as Shakespeare did. Many sold plays to whoever would buy them. Some had contracts to write a number of plays for a particular company over a stated number of years. At the end of the contract, or if the contract had not been fulfilled, the playwright might move on to another company. Collaborative writing was common, with writers sharing the money according to their contributions. The output was generally very high and the income relatively low, and even these small payments would normally be made in instalments as the play progressed. Playwrights could not always be relied upon to produce a steady turnover of new material. Many factors had to be considered by company managers: illness, spells in prison and sudden death were just some of the reasons playwrights might not meet their deadlines.

Robert Greene, of 'upstart crow' fame, was not the only playwright to die young. Christopher Marlowe died in 1593, a year after Greene. His was a much more violent death, but that wasn't all that surprising given Marlowe's violent life.

On 18 September 1589, Marlowe was involved in a fatal sword fight. An argument had started between him and an innkeeper's son named William Bradley. The cause of the argument is unknown but the two were fighting when Marlowe's friend, a fellow poet called Thomas Watson, happened to bump into them and drew his sword to part them. Marlowe withdrew from the fray but Bradley saw Watson with his sword drawn and 'leapt upon' him. Bradley cornered Watson, but the poet fought back and stabbed his sword six inches into Bradley's chest, killing him. Watson and Marlowe were both sent to Newgate prison charged with murder.* They were kept there for two weeks before they appeared at the

* Some historians state they were in fact sent to Finsbury prison, not Newgate.

Old Bailey. Neither was found guilty as it was believed they had acted in self-defence. Marlowe was discharged immediately but Watson remained in prison for almost five months until he received the Queen's pardon.

A few years later Marlowe was again in trouble with the law. This time, in May 1592, he had committed some unspecified violent offence in Shoreditch and was bound over to appear at the next General Sessions. In September that same year Marlowe fought with a tailor in his home town of Canterbury and again he was hauled in front of the authorities.

On 20 May 1593 Marlowe was arrested yet again, though not for a violent act on this occasion. Thomas Kyd, a playwright with whom Marlowe had once shared rooms, had given testimony (under pain of torture or threatened torture) that certain controversial documents that had been found among his possessions had in fact belonged to Marlowe. Marlowe was dragged before the authorities for questioning but released until a later trial date.

Ten days later Marlowe was in a boarding house in Deptford to meet with three rather dubious characters: Ingram Frizer, Nicholas Sheres and Robert Poley. The reason for their meeting is unknown but they spent the day together eating, walking and apparently getting along well enough, at least until after they had eaten supper. Then there was an argument over who should pay the bill. Marlowe suddenly pulled a dagger from Frizer's back pocket and attacked him, wounding him in the head, though not fatally. In the ensuing struggle Frizer appears to have wrenched his dagger back from Marlowe's hand and then stabbed Marlowe in the head, just above the eye, killing him instantly.

At least that is the official version. Marlowe may have been a spy for the Crown and there is speculation that he was in fact assassinated – his ex-roommate Thomas Kyd's testimony was merely a means to get to Marlowe.* Kyd himself died a year later at the age of 36. Perhaps the effects of torture had taken their toll. Regardless of how and why Marlowe met his end, the final outcome was the same. At the age of 29, after a short but eventful life, one of London's most promising playwrights, the author of *Tamburlaine*, *Doctor Faustus* and *Edward II*, was dead.

Shakespeare undoubtedly looked up to Marlowe and the two possibly collaborated on Shakespeare's *Henry VI* Part I.† Marlowe's influence continued long after his death. Shakespeare quite blatantly stole many lines from Marlowe's work, and even most of the plot of *The Jew of Malta*, which he turned into *The Merchant of Venice*. He may even have used Marlowe as inspiration for the character Mercutio, the witty and hot-tempered companion to Romeo in *Romeo and Juliet*. And several years after Marlowe's death Shakespeare is believed to have referenced his murder in *As You Like It*: 'it strikes a man more dead than a great reckoning in a little room'.

Marlowe's death was a great loss to English theatre and his absence must have been felt deeply by his fellow playwrights. Sadly it wasn't long before another of their

---

* It is far more likely that after a day of drinking, a fight simply got out of hand.

† The idea that it was in fact Christopher Marlowe who wrote the plays normally attributed to William Shakespeare has little evidence in fact. In order for this to be true, Marlowe would have had to fake his own death and remain out of the limelight for the next 20 years, allowing the name 'Shakespeare' to be given to his work.

number exited the scene. George Peele was enjoying some success in London in the 1580s and 1590s. He is believed to have collaborated with Shakespeare on the play *Titus Andronicus*, the most bloody of all the plays in the Shakespeare canon. The high gore content is often attributed to Peele's influence – he had also written the notably bloody *The Battle of Alcazar* (see Chapter 2). Although it is generally accepted that he wrote the whole of the first act and the first scenes of Acts II and IV of *Titus Andronicus*, Shakespeare and Peele likely collaborated closely throughout.

Outside of professional theatrical collaborations, Peele was good friends with Greene, Marlowe and Nashe. Like them he seems to have led an indulgent life. He died of the pox (syphilis), aged 40, and was buried on 9 November 1596. Of Peele's contemporaries, Shakespeare stands out for his longevity.

Shakespeare may appear to have had a rather sober disposition compared to his fellow playwrights. He, and his acting company, were certainly professionals in the world of theatre – they were reliable and not given to drunken brawls. But there are exceptions to every rule. In 1596, outside the Swan Theatre, William Wayte claimed to have been set upon by four assailants. In front of the Queen's Bench he swore that he had been in 'fear of death'. One of the accused was William Shakespeare. The playwright may simply have been in the wrong place at the wrong time when an existing dispute between Wayte and the other three accused got out of hand. However, those involved in the incident had links to the theatrical world, so Shakespeare may not have been an innocent bystander. The accused were ordered to post a bond (hand over an amount of money, similar to bail money) that would be forfeited if they broke the peace again.*

It wasn't Shakespeare's only brush with the law. In 1601 the Earl of Essex made a bid to overthrow the Queen. The afternoon before he and his conspirators marched towards the royal court, they had been entertained, by special request and with the promise of a grossly inflated fee, by the Lord Chamberlain's Men with a performance of *Richard II*, a play that depicts the overthrow and murder of a monarch.† Essex, and several others, were arrested and executed for treason, but the players seem not to have suffered at all.

⋆ ⋆ ⋆

Violence and political unrest were a common feature of late sixteenth-century England. Tempers were short, and most men carried weapons. It was the fashion that all gentlemen wore swords and virtually every adult male carried at least a dagger. Groups of men suddenly engaging in sword-fights with fatal results, as they do in *Romeo and Juliet*, was not just a convenient plot device; it was often the reality on London's streets. A notable spike in crime occurred in the 1590s. It was a tense decade generally in the capital, but 1595 was a particularly

---

* Shakespeare may have parodied the incident in *The Merry Wives of Windsor* and the character of Slender may be based on Wayte, but the evidence, like the character's name, is slender.

† Queen Elizabeth I was often compared to Richard II during her lifetime, and was doubtless annoyed to be associated with a king who was seen as weak and irresponsible.

tumultuous year, and one of the causes of disquiet was food, or the lack thereof.

Joyce Youings wrote that, in the late sixteenth century, 'Probably for the first time in Tudor England, large numbers of people in certain areas died of starvation.' Poor weather and an increasing population meant the demand for food often outstripped supply. London was growing at a fast rate and feeding the ever-increasing number of inhabitants was a considerable problem. The food shortages, among other grievances, prompted an estimated 35 riots between 1581 and 1602.

Things came to something of a head in 1595. There were 13 riots in June of that year alone. The pillories in Cheapside were torn down and a makeshift gallows was erected outside the Lord Mayor's house. On 29 June 1,000 apprentices marched to Tower Hill, where the city's gun-maker's shops were located. Order was eventually restored and the offenders severely punished – five of the apprentices' leaders were hanged, drawn and quartered. The city was placed under the equivalent of martial law, which included closing the theatres again, just as in plague years.

Even in times of relative plenty, many of the poor were malnourished. The wealthy, by contrast, were likely to be dyspeptic because of their rich diet, full of red meat and wine. There were constant problems with disease and the generally unsanitary conditions on London's streets, so most people probably felt ill most of the time, putting everyone in a bad mood.

Another factor was war. Conflict abroad, a constant feature of England in the 1590s, might be expected to remove the more violent members of society from the streets. They could enlist and be paid to commit acts of violence that, under other circumstances, would find them hanging at the end of a noose – 'There is no King … if it come to the arbitrement of swords, can try it out with all unspotted soldiers: some peradventure have on them the guilt of premeditated and contrived murder' (*Henry V*).

Problems occurred when troops returned from abroad. Pay was not always readily forthcoming while they were on campaign and stopped as soon as they returned to civilian life. Those who had been injured in the wars would have less chance of finding gainful employment at a time when vagrancy was common due to lack of work, and many would have been forced into criminality to support their families. There was no support mechanism available for those left damaged and traumatised by war.

Troops marching towards embarkation ports could also pose a problem. In the 1620s, another decade when crime rose dramatically, there were frequent complaints made to the Essex sheriff about disorderly troops, stragglers and deserters. The problem became so bad on one occasion that armed officers were stationed at crossroads to supervise the troops as they passed through.

Incredibly, despite all this ill-feeling and potential violence, more people in London died of accidental injuries than were killed by other people. Nevertheless, an audience member at one of Shakespeare's plays was likely to be very familiar with the sight of violence and death, whether accidental or intentional. Into this period of tension and unrest entered another playwright with a short temper and an attraction to controversy, Ben Jonson.

* * *

Jonson was born to a modest family and raised by his mother and step-father, a bricklayer. It might seem an inauspicious start but through the benefit of an excellent classical education, paid for by a family friend, and copious reading, Jonson became one of the most cultured men of the period. He did not attend university; instead he was apprenticed to a bricklayer, but abandoned the trade and volunteered for the army. He was particularly proud of the fact that he had killed a Spanish soldier whom he had challenged to single combat in the no man's land between the two armies. In the 1580s he left the wars on the continent and military life and returned to England to become an actor. By 1597 he was with the Lord Admiral's Men, but it soon became clear that his talents lay in writing rather than acting.

Jonson's writing career had a bit of a bumpy start. In 1597 he had collaborated with Thomas Nashe to produce the play *The Isle of Dogs*, a satire on the royal court. Whatever its literary merits, the play was not well received by the authorities because of its scandalous content. Nashe fled at the first sign of trouble but Jonson was arrested for 'Lewd and mutinous behaviour'. He was incarcerated in Marshalsea prison along with two actors from the play, Robert Shaw and Gabriel Spencer. The play was immediately suppressed and no copy has survived. Shaw and Spencer were released almost immediately but Jonson had to wait a few months to gain his liberty.

This early incident apparently did nothing to derail Jonson's writing career. The following year he produced his first great independent success, *Every Man in His Humour*, and more plays followed. But in September

1598 he was in trouble with the law again, and this time on the far more serious charge of murder.

Jonson had got into a disagreement with Gabriel Spencer, the same actor he was imprisoned with the previous year in Marshalsea. The cause of the argument is unknown but it escalated quickly, and the two men ended up in Hoxton Fields fighting a duel. Jonson later grumbled that Spencer had hurt him in the arm with a sword 10 inches longer than his own. He retaliated and killed Spencer, for which he was arrested. The incident caused considerable grief for the Lord Admiral's Men. Philip Henslowe, actor and theatre manager, complained that 'I have lost one of my company, which hurteth me greatly; that is Gabriel, for he is slain in Hogsdon [Hoxton] Fields by the hands of Benjamin Jonson, bricklayer'. The company was short one actor. The potential loss of a playwright seems to have been of less concern.

Jonson pleaded guilty to murder and was sentenced to be hanged, but he escaped the death penalty using a legal loophole known as 'benefit of clergy'. If a criminal could demonstrate that they could read, usually by reading a verse of the Psalms, he could claim to be a cleric and be tried in the ecclesiastical courts, which did not have the death penalty.

Benefit of clergy was not a simple 'get out of jail free card'. Jonson had all his possessions confiscated and he was branded on the thumb with the letter T for Tyburn. It marked him out for ever and meant that if he committed a second offence he could not escape the scaffold again.

Jonson clearly relished controversy and again caused consternation with the authorities in 1605 with *Eastward Ho!*, a play written in collaboration with George

Chapman and John Marston. The play ran into trouble when a Scottish courtier, Sir James Murray, complained to King James that Jonson had written 'something against the Scots'. Jonson was imprisoned with his collaborators, where they were threatened with having their ears and noses slit. The eloquent begging letters he wrote from prison eventually gained them all pardons.

Remarkably, given his penchant for getting into trouble, Jonson lived a long and highly acclaimed life writing plays, poems and masques (a form of theatrical entertainment consisting of dancing and acting performed by masked players) for King James's royal household. In 1628 an illness paralysed him and he was confined to his home for the remainder of his life. He died in 1637 at the grand age of 65.

★ ★ ★

After the deaths of Greene, Marlowe and Peele in relatively quick succession, Shakespeare enjoyed a few short years with very little in the way of competition. The arrival of Jonson on the theatrical scene in 1598 marked the start of a new era of prominent playwrights. Jonson, along with Thomas Dekker, Thomas Middleton and John Webster, all began to gain prominence towards the end of the sixteenth and the beginning of the seventeenth centuries. But Shakespeare's reputation was well established by then. Marlowe was best known for his tragedies and Jonson for his comedies, but Shakespeare was producing outstanding work across both these genres, as well as history plays – a breadth of talent in playwriting not matched by any of his contemporaries.

In other respects, he was rather narrow in his writing, producing only plays and poetry. Other writers were not only producing plays, they wrote poetry, masques and prose works too.

In his mature years Shakespeare was evidently successful, not just as a writer but as a businessman. He owned part shares in the acting company and used the profits from his work to invest in property back home in Stratford. He clearly felt confident enough in his success to apply for the title of 'gentleman', much to the derision of his fellow playwrights. The application had originally been made by Shakespeare's father John but had fallen into abeyance when John's prosperity also fell. William took up the cause and pushed it through the various channels to obtain the right to the title and a coat of arms. The design of the coat of arms, gold (or yellow), with a black banner bearing a silver spear, suggests Shakespeare may have been acting slightly tongue in cheek. Some have claimed the sight of the pompous Malvolio in *Twelfth Night* decked out in yellow stockings and black ribbons 'cross-gartered' is a reference to his coat of arms and a sign that he perhaps didn't take himself too seriously.

Despite the benefits that came with being a gentleman, playwriting was still a precarious profession. Shakespeare's company may have enjoyed the privilege and financial benefits of performing at the royal court, but on 2 February 1603 the Lord Chamberlain's Men gave their last performance before Queen Elizabeth. Soon after, the Queen fell ill and she died on 24 March, at the age of 69 (a remarkable lifespan given the amount of toxic lead she covered her face with every day – see Chapter 8).

Shakespeare's coat of arms by Sir William Dethick, Garter Principal King of Arms. The spear is a play on the family name.

Elizabeth was succeeded by King James VI of Scotland, known as James I in England. When the Queen died, the Lord Chamberlain's Men lost their patron and their company's name along with their monarch. Fortunately they were taken on by the new ruler and from then on Shakespeare was a member of the King's Men. But it would be a while before the newly named theatrical company could perform in public.

A new king and a new plague arrived almost simultaneously in London. The outbreak was so fierce that King James's coronation took place without any of the public attending the ceremony. All celebrations and processions were postponed until the following March (1604), when the rate of infection abated enough to allay fears of contagion from public gatherings. When the grand procession finally took place it must have been quite the celebration, as it marked not only the crowning of a new monarch, with all the associated pomp and pageantry, but also the end of the plague, or at least that particular outbreak.

The 1603 epidemic was the worst in 60 years, overshadowing even the 1592–3 outbreak. Over the hot summer the death rate in the capital soared to over 2,000 a week. In total it killed around one in five inhabitants of the city (estimates of the death toll vary between 25,000 and 38,000 people). Southwark, home to Shakespeare's Globe Theatre, was hit particularly hard, with estimates of 2,500 deaths in the parish in the space of just six months. Among the dead were two actors in Shakespeare's company, William Kemp and Thomas Pope.

The plague closed the theatres for the following seven years, with only occasional, brief re-openings. The King's Men were lucky; they were retained members of the royal household under King James and several payments were made without performance to enable them to survive.

There was little respite from the threat of untimely death. When one danger appeared to abate, another soon took its place. Food shortages struck again in 1608; that same year Shakespeare wrote the opening scenes of *Coriolanus*, where the angry citizens are protesting about food shortages. Inspiration might have been close at hand. Then the plague struck again and the theatres were closed for the summer and autumn, putting the players back on the road.

★ ★ ★

Shakespeare could avoid the worst of the plague by retreating to Stratford. And his income as a successful playwright, poet and property owner would have protected him against the worst of the food shortages. But the threat of violence was still rife. Even in later life, Shakespeare's association with the more temperamental playwrights of the age was not over.

Around this time he collaborated with George Wilkins on the play *Pericles*. Little is known of Wilkins's life apart from his brief career as a dramatist and his many brushes with the law, including one that involved Shakespeare. Wilkins's only known writings are from the period between 1606 and 1608. After that he left the theatrical life to run a London inn, which also doubled as a brothel.

Wilkins was a violent man, particularly towards women. In 1611 he was accused of 'kicking a woman on the belly which was then great with child', and a year later he was said to have 'outrageously beaten one Judith Walton and stamped upon her so that she was carried home in a chair'. He was also in court alongside Shakespeare, though neither was involved in a violent offence this time, but as witnesses to a marriage contract that had gone sour. Shakespeare had helped in the marriage negotiations, but after the wedding, arguments over money saw the newlyweds leave the family home to lodge with Wilkins.

In the last 10 years of his career Shakespeare collaborated often, not only with George Wilkins, but also with Thomas Middleton (*Measure for Measure* and *Timon of Athens*), Thomas Dekker (*Sir Thomas More*)* and John Fletcher (*Cardenio*, *Henry VIII* and *The Two Noble*

* This play is not usually included as part of Shakespeare's canon as his contribution is small.

Kinsmen).* He was perhaps acting in part as a coach or slowly handing over the reins to the new generation of playwrights.

With the exception of Wilkins, the younger playwrights appear to have led a more conventional and sedate life than their predecessors, avoiding violent excesses. Thomas Middleton seems to have had a relatively trouble-free existence and died at his London home aged 47. Dekker had a lifelong problem with debt causing him to spend more than one spell in prison. In 1612 he was imprisoned for a £40 debt and remained incarcerated for the next seven years, still writing from inside the jail, until he was able to free himself. Fletcher had a prolific literary output, both as sole writer and in collaborative works. His chief writing companion was Francis Beaumont with whom he shared not only writing credits but also his home, holidays, clothes and mistress.† Beaumont moved out when he married but Fletcher died a bachelor at the age of 46 from the plague.

★ ★ ★

From 1610, Shakespeare's creative output, never as considerable as some of his contemporary playwrights, slowed, and by 1613 it appears to have stopped completely. The writer, perhaps tired of London life, retired to the countryside and returned to his native

* *Cardenio*, a play apparently based on Don Quixote, has been lost. *The Two Noble Kinsmen* has only been included in Shakespeare's list of works in recent years. It was probably not included in the first collection of his work because Fletcher was still living when it was published.

† The exact living arrangements aren't known.

Stratford. Poor health may also have been a contributing factor in the decline of the Bard's output and a reason to return to the healthier environs of rural England.

Shakespeare may have written about many deaths in his plays and poetry but his own death, like most of his life, is shrouded in mystery. He had clearly been feeling unwell in January of 1616, when he called for his lawyer, Francis Collins, and asked him to draft his will. He must have made a recovery as the document was not completed. But on 25 March, Collins returned, the will was finished and Shakespeare signed each page with his own trembling hand. He died a month later on 23 April, the day that might also have been his 52nd birthday.

The actual cause of death is unknown, but that hasn't stopped anyone from speculating. One possibility comes from the diary of John Ward: 'Shakespeare, Drayton and Ben Jonson had a merry meeting, and it seems drank too hard, for Shakespeare died of a fever there contracted.' Ward was writing his anecdote 50 years after the event and it is unlikely to be accurate. Another unreliable report claims 'he caught his death through leaving his bed when ill, because some of his old friends had called on him'. Other theories include syphilis, alcoholism, typhoid and influenza.

The previous winter had seen an epidemic of influenza. The typhoid theory also has some credibility, in part due to a report from a seventeenth-century doctor who noted that fevers were 'especially prevalent in Stratford and that 1616 was a particularly unhealthy year'. A rivulet ran past the playwright's home, New Place, and it was later shown that these small streams were carriers of the typhoid bacteria. The four weeks between signing his will and his death would be about right for a case of typhoid fever. A death from something contagious is also suggested by the very prompt burial of Shakespeare's body. He is said to have been buried at a depth of 17 feet, which seems extraordinarily deep, even if out of fear of contagion.

Whatever afflicted the dying poet, he was almost certainly attended by John Hall, a doctor who had married William's daughter Susanna in 1607. Hall had a thriving medical practice in Stratford even though it seems he never took a medical degree, or if he did no record survives. He was never licensed by the College of Physicians, nor did he have a bishop's licence to practise, but this is not particularly unusual for a time when the profession was poorly regulated. Despite his apparent lack of qualifications he seems to have been highly regarded by his patients.

Hall kept casebooks but, unfortunately, they contain no details of his father-in-law's illness. The earliest records date from 1611, but all tended to detail his successes rather than cases where his patient succumbed. For this reason, even if other casebooks by Hall exist and were to be found, it is unlikely that the details of William Shakespeare's fatal illness would appear in them.

Whatever brought him there, on 25 April 1616 William Shakespeare was interred in the chancel of the church where he was baptised, Holy Trinity, Stratford. He left behind his wife and two daughters, along with 39 plays and over 150 short and long poems, many of them considered to be some of the best ever written in the English language.

CHAPTER TWO

All the World's a Stage

The play's the thing

Hamlet, Act 2, Scene 2

When Shakespeare arrived in London, probably in the late 1580s, it was the beginning of a new era in theatre, and performances were much closer to the art form we might recognise today than in the previous years. Dedicated playhouses were being built in the capital and the companies putting on performances were increasingly professional. New genres, venues and performance styles were being tested out on audiences hungry for entertainment.

Theatre-goers in Renaissance London were not shy in letting their opinions be known. Plays could draw enormous crowds but they were repeated only as long as audience numbers stayed high. A staggering number of plays only saw one production before disappearing for ever. It was an experimental time, but playwrights soon learned to give the audience what it wanted – fantastical stories, fight scenes and bloody deaths. It was a very different world to the one we know now, but this environment produced some of the greatest theatrical works ever written.

★ ★ ★

Shakespeare's first theatrical home, the imaginatively named 'Theatre', was only the second purpose-built playhouse to be constructed. It was built in 1576 in Shoreditch, just outside London's city walls and the reach of the City authorities. The venue only closed after an argument with the landlord over the lease. Fortunately for the Theatre's owner, actor-manager James Burbage, the lease only applied to the land. When negotiations failed, Burbage had the building dismantled and moved south of the Thames to Southwark, where it was resurrected as the Globe in 1599.

South of the river, still out of the reach of the City authorities, was a place where Elizabethan Londoners went to have fun. Southwark was the area where theatres were generally located, among public gardens, bear-baiting arenas and taverns. It was also home to over 100 brothels, or 'stews' as they were known.*

---

* There are two possible explanations for brothels being referred to as stews. It may be because of their location near ponds used to breed fish for the dinner table; it may also be a connection to Roman bath-houses, often associated with prostitution, which contained a sweating room and therefore a stove, which in Norman French is called 'estues' or 'estuwes'.

There was a particularly close association between these houses of ill repute and their neighbours, the playhouses. The Rose Theatre had been built on the site of a former brothel, hence its name, the Elizabethan slang for a prostitute. The business partners Alleyn and Henslowe, who owned the Rose Theatre, also owned a number of stews in the area. Southwark, therefore, had something of a reputation and unsurprisingly bear-baiting and brothels, and their associated dangers from maulings and venereal disease respectively, are frequently alluded to in Renaissance plays. A civic edict ordered wherrymen (ferrymen) to moor their boats on the northern bank of the Thames at night, so that 'thieves and other misdoers shall not be carried' to the brothels and taverns of Southwark.

Ideas of what was considered entertaining were very different 400 years ago. For example, in the morning, the average Londoner might go to Tyburn, or to a number of other sites of execution in and around the capital, to watch a hanging. They could then walk across London Bridge, the only bridge crossing the Thames at the time, towards Southwark. At the south end of the bridge they would pass under the Great Stone Gate, where the decapitated heads of criminals and traitors were prominently displayed on pikes (one visitor to London in 1592 counted 34 heads on display). In the afternoon they could watch similar scenes acted out in theatres where characters would be dragged off to be executed and fake heads would be brought out onstage to show the deed had been done.

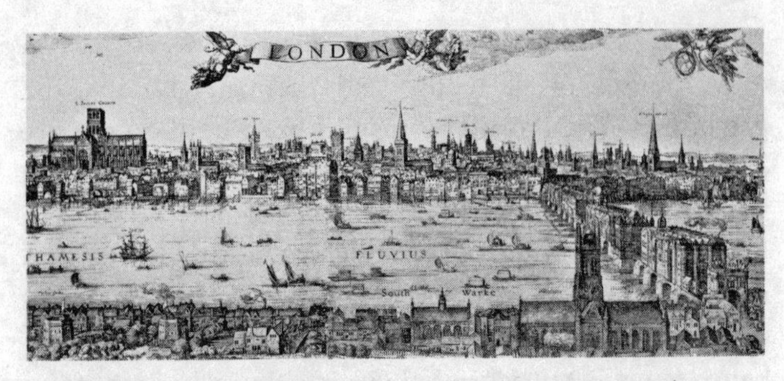

Panorama of London produced in 1616 by Claes Jansz Visscher showing the Globe and the Bear Garden in the bottom left and, in the bottom right, the heads of traitors above the Great Stone Gate on the south side of London Bridge.

* * *

The popularity of theatres grew rapidly in the 1590s, when Shakespeare was just starting to make his presence felt on the theatrical scene. There were a handful of purpose-built playhouses in operation around the capital competing for audiences and more were constructed over the following decades. Knowledge of what these buildings looked like and how they operated has been pieced together from evidence from a variety of sources including archaeological excavation, contemporary letters and diaries as well as sketches and maps; there are even hints found in the plays themselves. In the opening of *Henry V*, Shakespeare describes the surroundings as 'this wooden O' and asks the audience to imagine the scene at the Battle of Agincourt instead of a bare stage in a London suburb. The actor speaking these words for the first time was standing in front of an audience either at the Curtain Theatre in Shoreditch or the Globe on the South Bank.

These early theatres were built from wood in a roughly circular or polygonal shape, which was as close to a circle as the materials and Elizabethan building techniques allowed. In the centre of the building was the pit or yard, open to the elements, where audience members (called groundlings) stood to watch performances in front of a raised stage. The stage was covered by a canopy richly decorated in bright colours and gilding, 'this majestical

The interior of the Swan Theatre sketched by Johannes De Witt in 1595.

roof fretted with golden fire' (*Hamlet*).* Encircling the yard were three storeys of covered galleries.

The theatres attracted their audiences from all parts of society, from the richest to the poorest. Within the walls of the theatre, nobility rubbed shoulders with ne'er-do-wells. Prostitutes and pickpockets plied their trade among the jostling crowds (though both professions might have found an easier living in the nearby taverns). Ticket prices were extremely cheap: for one penny anyone could stand in the yard under the open sky to watch the performance.† Seats under cover in the galleried areas were more expensive and prices rose as comfort and proximity to the stage increased.

The amphitheatres were open year-round, apart from regular closures for 37 days during Lent, and eight weeks during the summer for touring. Performances, plague permitting, continued through autumn and winter. It was a determined and hardy audience member who could stand through several hours of a performance in the freezing cold or pouring rain.

England experienced a dramatic change in climate around the turn of the seventeenth century. Shakespeare seems to have made the most of the unpredictable weather and open-air conditions in his many references to storms (*Pericles*, *Twelfth Night* and, of course, *The Tempest*). His audience could well have been standing

---

* This canopy, supported by two enormous pillars, was referred to as 'the heavens' and the area below stage, accessed by a trapdoor in the middle of the stage, was consequently known as 'hell'.

† For comparison, one penny could buy a loaf of bread. Actors, and artisans who had completed their apprenticeship, earned around a shilling a day (12 pence).

in a torrential downpour while hearing the lines from *Julius Caesar*, 'Why, now, blow wind, swell billow and swim bark! / The storm is up, and all is on the hazard.' When Hamlet complains, 'The air bites shrewdly, it is very cold', he wasn't just setting the scene, it really was freezing. Europe and North America were in the middle of the 'Little Ice Age'. The seventeenth century was the coldest of the entire second millennium and 1601, the year *Hamlet* was probably first performed, was its coldest year.

Nevertheless, people don't seem to have been put off by the inclement weather. Theatres regularly attracted audiences in their thousands. Johannes de Witt, a Dutch scholar who visited London in 1595–1596, estimated that the Swan playhouse could hold 3,000 people. Nearby, Shakespeare's Globe had a capacity of at least 3,000, according to the Spanish ambassador, who went to the rebuilt theatre in 1624; other estimates have it at a capacity of 3,300. It would have been crowded, to say the least. Perhaps the close proximity of 3,299 others helped to keep you warm.*

A trip to the theatre would have been an assault on all the senses. A combination of foul smells from the nearby polluted river, bear-baiting arenas and the neighbourhood breweries and tanneries (known collectively as the 'stink trades') would have pervaded the air. Inside the theatre food sellers weaved through the crowd offering nuts, meat and shellfish. Bad breath and body odour, from

---

* In the new Globe Theatre, opened in 1997 close to the original site of Shakespeare's 1599 theatre, capacity was more than halved to 1,500 owing to health and safety considerations.

thousands of unwashed audience members huddled together with no toilet facilities, created 'a foul and pestilent congregation of vapours' as Shakespeare put it in *Hamlet*. Men could find a convenient pillar or wall; women didn't even have to move and could take advantage of their long skirts to relieve themselves on the floor where they stood. This was nothing unusual.

Surprisingly, urine was a valuable commodity because of its high urea content (around 5 per cent in a fresh sample; the rest is mostly water). Over time chemical changes can convert the urea to ammonia and its high pH means it can break down organic material. Tanners would soak animal hides in stale urine to soften them. Ammonia, via nitrates, could also be used to produce saltpetre, a key ingredient in gunpowder, and a gunpowder manufacturer obtained the rights to scrape the floor in Ely cathedral where the women stood during services. No testing for urine was done on the site of the Rose Theatre when it was excavated in 1989, but if women were uninhibited in a place of worship, it seems unlikely they would have restrained themselves at the theatre. In *Twelfth Night* Duke Orsino may have been right when he said that the women in his audience 'lack retention'. The lack of a roof at the Globe, and other theatres, may have made it bitterly cold in winter but at least it was well ventilated.

Then there was the noise. Productions in the amphitheatres had to be loud in order to be heard across the open space. Trumpets blared to herald battles and fireworks and pyrotechnics were used enthusiastically to illustrate cannon-fire and lightning storms, despite the obvious dangers of sparks in a densely packed and highly combustible venue. And the noise wasn't just coming from the stage. Elizabethan and Jacobean audiences could be, indeed were expected to be, vocal during the performance.

★ ★ ★

Going to the theatre could be a considerable hazard; health and safety considerations were not paramount in Shakespeare's day. On 29 June 1613, the Globe Theatre burned to the ground during a production of *Henry VIII* (known at the time as *All is True*). Incredibly, given the number of people likely to have been squashed into the building at the time, no one died.* An account of the event from Sir Henry Wotton, an English author and diplomat, shows how lucky they were:

> Now, let matters of state sleep, I will entertain you at the present with what has happened this week at the Bankside. The King's players had a new play, called *All is True*, representing some principal pieces of the reign of Henry VIII, which was set forth with many an extraordinary circumstance of pomp and majesty, even to the matting of the stage; the Knights of the Order with their Georges and Garter, the Guards with their embroidered coats, and the like: sufficient within a while to make greatness very familiar, if not ridiculous. Now, King Henry making a masque at the Cardinal Wolsey's house, and certain chambers being shot off at his entry,

---

* Perhaps wary of its fiery history, it was not until 13 years after its opening that the new Globe Theatre was brave enough to stage a production of this play.

> some of the paper or other stuff wherewith one of them was stopped did light on the thatch, where being thought at first but idle smoke, and their eyes more attentive to the show, it kindled inwardly and ran round like a train, consuming within less than an hour the whole house to the very grounds. This was the fatal period of that virtuous fabric, wherein yet nothing did perish but wood and straw and a few forsaken cloaks; only one man had his breeches set on fire, that would perhaps have broiled him if he had not by the benefit of a provident wit put it out with bottle ale.

Audience members were not always so lucky. On 13 January 1583 a huge crowd gathered to watch the bear-baiting at the Paris Garden in Southwark. Mounted up on rickety scaffolding, the building suddenly collapsed, wounding 200 to 300 spectators, some severely, and killing seven. This was nothing new or particularly surprising as there had been similar fatal collapses before this. What made this incident the more shocking was that it happened on a Sunday. The more conservative members of society saw it as a judgement from God and used it as an argument against all forms of entertainment that they saw as licentious.

It wasn't just poor building work that threatened the welfare of Elizabethan audiences; the plays themselves sometimes posed risks. In 1587, the Lord Admiral's Men were performing in London, probably at the Theatre playhouse. One of the actors was tied to a pillar and an inexplicably loaded musket was fired at him. It 'missed the fellow he aimed at and killed a child, and a woman great with child forthwith, and hit another man in the head very sore'. The Lord Admiral's Men went into a tactical temporary retirement and disappeared from the records for over a year.

★ ★ ★

It's not difficult to see why City authorities and Puritans had such a low regard for theatres. When Shakespeare's company tried to expand their operations to an indoor theatre north of the river, it was resisted for years. Finally, in 1608, they were granted permission to perform in an indoor playhouse at Blackfriars.* It might have been seen as a step up in the theatrical world, as its location, north of the river, was more respectable; the audience, all seated, was considerably smaller, and ticket prices were much higher. But the company didn't turn its back on the Globe. Instead, they based themselves at their indoor venue for the cold winter months but returned to their South Bank home during the summer.

The Blackfriars venue was considerably smaller than the Globe and productions had to be adapted to the new space. Trumpet calls could be deafening in the enclosed room and were abandoned for quieter instruments. Extensive use of pyrotechnics inside could choke an audience and make the stage difficult to see through the smoke. Light entered, and presumably smells and smoke escaped, through small windows located near the high ceiling and extra lighting was provided by candles.

Performances had to be adapted for the more claustrophobic environment, but some plays were better

---

* From a historical point of view the Blackfriars Theatre is the more important venue as it formed the basis for all subsequent indoor theatres.

suited than others. *Macbeth*, a play set mostly at night that includes witches, ghosts and murders, would have been quite a different experience in the gloomy setting of Blackfriars compared to the bright open space of the Globe. The smaller enclosed space allowed for more intimate scenes and elaborate special effects to be incorporated into plays, but battle scenes had to be scaled down to fit on the smaller stage. In those days actors fought with real swords and audience members could even sit onstage to watch the action up close. In 1622, at the indoor Red Bull Theatre, one audience member got too close. A felt-maker's apprentice was accidentally injured by one of the actors, Richard Baxter, during a particularly exuberant sword display. The very real risk of injury must have made for an exhilarating experience for some.

Audiences at Blackfriars might have been expected to be more genteel, but things could still get pretty raucous. On one occasion an Irish lord blocked the view of the Countess of Essex at the Blackfriars Theatre. An argument ensued between the lord and the Countess's escort that quickly developed into a duel. Later in the seventeenth century, during a performance of *Macbeth* at the Blackfriars venue, a nobleman perched on the edge of the stage spotted someone he knew entering the theatre on the other side. He stood up and walked through the action onstage to greet his friend. When one of the actors rebuked the nobleman he was slapped for his impudence and the audience rioted.

★ ★ ★

Going to the theatre was obviously rather different in Shakespeare's day, more crowded, noisier, undoubtedly smellier, and a rather more risky experience than now. None of this seems to have deterred Elizabethan audiences. Even though there were only a handful of venues, it is estimated that in 1595 around 15,000 people went to the theatre every week. The appetite for something new to watch was difficult to sate.

To keep the crowds interested, the turnover of plays was considerable and productions were only repeated if they were popular. For example, in the 1594/5 season the Lord Admiral's Men enjoyed a run of 49 weeks in the Rose Theatre, interrupted only by Lent and a short break for essential repairs. In that time they put on 273 performances of 38 different plays, 21 of which were completely new. Only eight of those plays were performed the following season. Playwrights were churning out new material at an incredible rate and rehearsal time was severely limited. An Elizabethan actor's capacity for learning new lines, and retaining them to build up a considerable repertoire, must have been phenomenal.

Modern audiences can easily miss connections between plays because they are usually seen in isolation. Frequent visits to the theatre and a rapid turnover of material meant an Elizabethan audience was much better placed to pick up on links between plays as well as subtle references to work by other playwrights. Writers like Shakespeare could also afford to spread the narrative over several plays, such as in the three parts of *Henry VI*, because they would be presented in relatively quick succession.

But there were other constraints on playwrights that just don't exist today. Performances started at 2 p.m. and were restricted to a mere 'two hours' traffic of our stage' (*Romeo and Juliet*), by order of the civic authorities, although this doesn't appear to have been strictly enforced. Even so, plays had to be over before dark in the open-air theatres (fine in the summer, but it could mean a 4 p.m. finish in an English winter, at the latest). Most performances appear to have lasted between two and three hours, including a jig at the end.

A playwright may have submitted a beautifully crafted, elegantly worded script full of poetical speeches and witty dialogue, but changes would be inevitable in performance. Manuscripts would be annotated with stage directions and edits made to keep everything within the restricted time. Stage performances would have focused on the action and kept the minimum of explanation to make the plot comprehensible. *Henry V*, for example, was probably staged without many of the stirring speeches that are favourites with modern audiences.

The different versions of Shakespeare's plays that have come down to us offer possible insights into how plays were edited for performance, and how they were amended over the years to bring new life to old plays in revival. For example, the quarto version* of *Henry V* was printed shortly after the play first appeared on the stage and the text was probably taken from the transcript of a performance. It is much shorter than the Folio

* Around half of Shakespeare's plays were published individually during his lifetime as quarto editions, so called because the paper was folded twice to make four leaves.

version,* which most likely came from the original manuscript, and captures all of Shakespeare's poetic vision for the play. The full text is a more enjoyable read for us today than the truncated version put onstage.

Tastes in drama were also different. For example, *Pericles*, rarely performed today, was a huge hit with Shakespeare's audiences. The popularity of *Titus Andronicus*, another play that eventually fell out of favour for centuries owing to its violent nature, was still drawing envious remarks from Ben Jonson 20 years after it was first performed. By contrast, *Anthony and Cleopatra* was not nearly so popular, and would have been lost for ever if it had not been included in the First Folio. But by and large, Renaissance audiences went to see the same Shakespeare plays that still pull in the crowds over 400 years later. This raises the question, how different were the performances themselves?

One difference was the style of acting, which would probably have been very formal and stilted to modern eyes. In Elizabethan times, actors had been expected, by and large, to stand still and address their lines to the audience using specific hand gestures to express emotion and sentiment. But performers were increasingly using a more lifelike style that often brought praise from critics. Shakespeare's company was certainly moving towards a more naturalistic mode of bringing their characters to life onstage. The playwright even made fun of how some

---

* After his death, John Heminges and Henry Condell, actors in Shakespeare's company, collected together all of his works from the author's own copies or the company's manuscripts, and compiled them into one volume, known as the First Folio.

actors still used a very formal style, as in *Hamlet* when the young prince talks to the troupe of actors just arrived at Elsinore Castle:

> Speak the speech, I pray you, as I pronounced it to you, trippingly on the tongue; but if you mouth it, as many of your players do, I had as lief the town-crier spoke my lines. Nor do not saw the air too much with your hand, thus, but use all gently; for in the very torrent, tempest, and, as I may say, the whirlwind of passion, you must acquire and beget a temperance that may give it smoothness.

Set design was also very different. No curtain could be brought down to hide complex set changes, and so simple things like a throne would signify the setting was a royal palace. Signs might be placed over doors to show whose home or which tavern they signified, and black hangings might be used to identify the play as a tragedy. Lighting couldn't be controlled in the open-air theatres and so actors holding torches or lanterns showed it was night-time. A prologue or chorus was often used to set the scene and describe sudden changes in setting. Mostly the stage was fairly bare, 'a sterile promontory' as Hamlet put it, but this certainly wouldn't have detracted from the experience. Elizabethan audiences went to hear plays rather than see them, and just because there wasn't much in the way of set design didn't mean there was nothing of interest to look at on the stage.

Money might have been saved on the sets but considerable investments were made on clothing and costumes. Those in the higher echelons of society, mainly knights and nobles, had a custom of bequeathing their clothes to their servants. These servants couldn't have worn the clothes left to them as there was a strict etiquette about who could wear what in Elizabethan England, so the servants would sell the clothes on to acting companies. This meant actors could be clothed in exquisite garments bought at a knock-down price. Even elaborate clerical garments could be easily obtained thanks to the Reformation.

Costuming choices seem to have been made based on appearance rather than accuracy. A sketch of a production of *Titus Andronicus* from 1595 shows half the cast in Roman garb and the rest in contemporary Elizabethan clothes. This was probably not due to a restricted wardrobe budget but rather a deliberate choice.

Elizabethan theatre was not about realism, faithful reproductions of historical events or moral lessons – it was about entertainment. Historical accuracy was only a

Sketch apparently of a performance of Titus Andronicus *made by Henry Peacham in 1595. It shows Tamora pleading with Titus. Behind her are her two sons with their hands tied and behind them stands Aaron.*

minor consideration and chronology suffered particularly under Shakespeare's quill. He could skip over several years in a few lines and travel backwards in time from one scene to the next. The Bard was perfectly happy to change events, omit certain characters or introduce new fictional ones into his history plays when it suited him.

There are many other things within Shakespeare's plays that today seem anachronistic. Clocks strike when they shouldn't. Cleopatra plays billiards centuries before the game, or anything like it, was invented. Macbeth talks of payments made in dollars before such a currency existed. In short, considerable artistic licence was in use.

A playwright's first consideration was the drama and spectacle of a play, but practical issues of staging were a vital consideration. After performing at public theatres, acting companies might have to travel to a much smaller venue for a private performance paid for by a rich patron. Props, costumes and staging would be kept to the very minimum for easy transport. The staging of the play would have been altered on the fly depending on what space was available to them. For example, without a trapdoor available, the gravedigger's scene in *Hamlet* is likely to have been cut.

On large stages like the Globe, plays could be written specifically for the facilities available. But even when the space was large and prop stores were close to hand, Shakespeare kept things to a minimum. Small props, such as letters and rings, might have been essential to the plot but it is estimated that 80 per cent of the scenes he wrote for the Globe require no props whatsoever. However, he did make exceptions for some of the gorier scenes when body parts and blood were used in abundance to appeal to the Elizabethan and Jacobean audiences' tastes.

★ ★ ★

Shakespeare referred to the stage as 'this unworthy scaffold' (*Henry V*) – not only a show of modesty but also a comparison with sites of execution. Shakespeare and his fellow playwrights were trying to attract the same audiences that went to watch bear-baiting and fencing displays for the same price as a theatre ticket, and also had to remember that the public could witness floggings, dismemberments and executions of criminals for free. In terms of spectacle, theatrical performances had to give value for money. Shakespeare scattered violence, blood and gore liberally throughout his plays. His characters fight each other with swords, some are stabbed or decapitated, hands are sliced off, tongues are cut out and eyes are gouged.

Perhaps the most iconic image people associate with Shakespeare is also its most macabre: Hamlet holding a human skull. It might have been relatively easy to bribe someone to steal a skull from a local charnel house but, despite Elizabethan familiarity with death and decapitated heads on spikes, using genuine human remains in a play might have been seen as disrespectful. Nevertheless, several productions of Hamlet have used real skulls. In 1982 the pianist André Tchaikowsky died and bequeathed his skull to the Royal Shakespeare Company for use in *Hamlet*'s graveyard scene. Though it was often used in rehearsals, actors were unnerved by the skull and it wasn't used in a live performance until 2008.

To modern audiences Shakespeare may seem quite macabre in his tastes, but he was certainly not exceptional among his contemporaries. Compared to John Webster's *The Duchess of Malfi*, or *The Battle of Alcazar* by George Peele (a play with a 'bloody banquet' that features 'dead men's heads in dishes' and that later calls for three characters to be disembowelled onstage), Shakespeare was usually quite restrained in his depiction of bloody and violent acts.

Blood and gore on the Renaissance stage was the norm, and directions for characters to be wounded, hurt or stabbed feature in more than 150 contemporary plays. As witnesses to violence and death on an almost daily basis, Elizabethan theatre audiences would not have been easily shocked. They were also in a good position to be able to judge the accuracy of many of the gorier stage props used, including the severed hands and tongues in *Titus Andronicus* and the decapitated heads brought onstage in several of the history plays – *Henry VI* Part II calls for three in the same play! The real thing was visible on spikes not far from the theatre.

Stage make-up and props were available, but how realistic these might have been is not so easy to judge. Neither is the effect a playwright was trying to create. A suggestion of how artificial body parts might be made for the stage comes from Webster's *The Duchess of Malfi*. Severed hands are presented to the Duchess but it is a trick. As Ferdinand explains,

> Excellent, as I would wish; she's plagued in art.
> These presentations are but framed in wax
> By the curious master in that quality,

Vincentio Lauriola, and she takes them
For true substantial bodies.

In the sixteenth and seventeenth centuries wax effigies were sometimes produced of recently deceased nobility. A particularly famous one of Prince Henry, who had died on 6 November 1612, was drawn through the streets of London – the same year that *The Duchess of Malfi* was first performed at the Blackfriars Theatre. Particular attention was paid to the face and hands to make them resemble the Prince as closely as possible. Funeral effigies were elaborate constructions as well as realistic – Prince Henry's wax figure could sit and stand. It was common practice to name the artist who had made these impressive figures and so Webster was following tradition when he names Vincentio Lauriola, even though no such person is known to have existed. In *The Winter's Tale* Shakespeare has a statue of Hermione brought onstage as a memorial to the character that died earlier in the play. The audience is told the uncannily lifelike effigy has been newly produced by the Italian master Julio Romano, who was a real-life Italian Renaissance artist.

As well as direct references to body parts in play texts and stage directions, a list of props and costumes belonging to the Lord Admiral's Men has survived. Their collection of theatrical possessions included 'Mahemetes head', 'Arogosse head', 'iii Turkes heads' as well as numerous limbs and a wooden leg.

Perhaps even more gruesome than decapitated heads, individual eyes are occasionally required to show to the audience. Today lychees are often employed when actors need to be seen to have an eyeball gouged out onstage, as the size, colour and texture of the fruit make excellent eye substitutes. Although the fruit is referred to in its native China as far back as 2000 BC, the first mention of it in the West is in 1646, long after Shakespeare was dead and buried. Elizabethans probably didn't need to go to such lengths to produce fake eyeballs when the real thing was readily available from nearby slaughterhouses.

Another difficulty was blood. There are many references to blood in plays of this era and several blood substitutes are known to have been used to create a good effect. Red vinegar and wine could be stored in bladders or soaked into sponges and concealed about an actor's person. Then at the critical moment he would squeeze the sponge or pierce the bladder to make the 'blood' flow. One problem with the use of vinegar or wine is that, though they would look fairly convincing in a puddle onstage, they don't adhere well to the skin. Something stickier, and with a more intense colour, was needed for daubing on actors or props.

Several references have been found to the use of vermilion, a mercury compound of sulfur that makes a bright red pigment. One such reference is in the payments made by Canterbury officials in 1528–9, apparently for a play on the subject of St Thomas Becket. The pigment was presumably used to illustrate blood at the moment of the saint's martyrdom.* Vermilion is a very distinctive red but it is not a particularly good match for blood. Its striking colour would enable an audience to easily

* Vermilion (also known as cinnabar) was a versatile pigment that could be used as paint, ink and even blusher.

understand that it represented blood in certain situations, but sometimes it just wasn't good enough and a play called for the use of the real thing.

Animal blood from pigs, cattle or sheep would have been relatively easy to obtain from nearby abattoirs or butchers, but not all blood behaves in exactly the same way. Clots are formed when blood is exposed to the air, a blood vessel is damaged or blood flow ceases and the blood begins to pool. These clots are made of platelets (fragments of red blood cells) and a mesh of long fibres called fibrin. The fibres clump together to form a net that traps more platelets and other cells in the blood. Clots are vital to prevent blood loss and to stop bacteria and other sources of infection from entering a wound, but this also means that liquid blood, left exposed to the air, doesn't stay liquid for long, at least in small quantities (seconds to minutes). Large volumes of blood can take a long time to fully clot, hours even. Compared with human blood, large quantities of cow's blood clot more slowly and remain liquid for longer before turning into a sticky jellied mess that won't flow. Sheep's blood takes even longer.

Using sheep's blood meant there was plenty of time to collect it from a nearby abattoir and store it in bladders until it was needed onstage. When the moment arrived to pierce the bladder and let the blood flow, it was most likely to remain liquid. This fact was well known to the Elizabethan actors and entertainers, even if they knew nothing of the science of clotting.

Reginald Scot, in his 1584 book *The Discoverie of Witchcraft*, describes a trick that could be performed to allow someone to appear to stab themselves. The trick called for the use of 'a gut or bladder of blood' placed between a 'plate' and a 'false belly', 'which blood must be of calf or of a sheep; but in no wise of an ox or cow, for that will be too thick'. Although Scot refers to the thickness of the blood, a sheep's blood is actually no 'thinner' than a cow's; the red blood cell mass (in healthy animals) is the same. Scot probably meant how quickly it stopped flowing, or clotted, and just didn't have the technical word for it.

Bladders of sheep's blood would therefore have been a fairly standard prop in an Elizabethan theatre. But sometimes even real blood wasn't gory enough and stage directions became even more graphic. One play contains a note simply calling for 'raw flesh'. Annotations in the margins of Peele's *The Battle of Alcazar* detail how three characters could appear to be disembowelled in front of the audience. The notes call for '3 vials of blood & a sheeps gather'. The gather was a bladder holding the liver, heart and lungs of the sheep. A small flask of its blood was given to each actor to burst open at the appropriate moment.

Regardless of what substance was used, all of these blood types and blood substitutes would stain if they came into contact with the costumes. Clothing was the most valuable part of an acting company's possessions, but laundering the delicate material used for the ornate costumes of the tragedies would have been difficult, especially given the limited cleaning materials available at the time. One option was soap, which is excellent for removing greasy deposits but, depending on the composition of the soap, probably wasn't good for removing blood stains.

Another option was lye, which can be made from the ashes of hardwood. During their life these trees accumulate potassium, which doesn't burn and so is left concentrated in the ashes. When mixed with water the potassium forms potassium hydroxide, or caustic potash, a corrosive liquid that destroys organic material such as blood. However, if the mixture is too strong it can also corrode the fabric and harm the hands of the person doing the laundry.*

A slightly less caustic, but possibly more unpleasant, approach would be to wash the clothes in stale urine. The ammonia formed from the urea would also break down organic matter that causes blood stains but it is much less corrosive than lye, so the fabric of the clothes was less likely to be damaged. However, the dyes used to colour Elizabethan fabrics would not have been very colour-fast and repeated washing would fade the costumes, so it was to be avoided if at all possible. Blood onstage was therefore carefully controlled.

The 1649 play *The Rebellion of Naples*, by an author known only as T. B., calls for an on-stage decapitation and even suggests how it might be done. 'He thrusts out his head, and they cut off a false head made of a bladder fill'd with blood.' Surely blood spurting uncontrollably all over the stage would have theatre producers wincing at the potential laundry bill. But in 1649 the theatres were closed, and had been for seven years thanks to the Puritans, so the play was probably written as a closet play (to be read rather than performed) and these notes were

* Lye can be so corrosive that it has been used to dissolve the bodies of murder victims.

merely suggestions rather than practical details on how it should be staged.

In plays that were performed onstage, deaths that were more difficult to act out might take place offstage. Characters returning from a violent episode that took place offstage could appear in front of the crowd with themselves and/or their props already bloodied. But if a character was to be wounded or killed onstage, the directions were often quite specific. For example, in the King's Men's production of *The Princess* performed around 1637: 'Bragadine shoots, Virgil puts his hand to his eye, with a bloody sponge and the blood runs down.'

In the case of Shakespeare's *Julius Caesar*, the script calls for Caesar to be killed onstage by eight conspirators.* There are references to blood, and the 33 stab wounds specified in the text are likely to make one hell of a mess, but all this can be carefully orchestrated.

At the moment of the attack, Caesar is surrounded by the conspirators; it looks dramatic and also shields the audience from what is actually going on. It would be enough for the audience to see swords drawn and Caesar collapse to the floor to understand that he had been stabbed many times. A pool of blood can then be made on the stage and a pre-bloodied mantle can be laid over him. Everything can be carefully controlled to avoid ruining not just Caesar's costume but those of his murderers too.

---

* According to Plutarch's *Lives*, Shakespeare's source for the play, the conspirators pressed together so eagerly to stab Julius Caesar that they wounded each other.

After the deed is done Brutus calls on his fellow conspirators to 'Stoop, Romans, stoop, / And let us bathe our hands in Caesar's blood / Up to the elbows and besmear our swords'. Carefully applying the blood from a pool on the floor means staining the costumes is less likely. It is also important in terms of plot, as when Mark Antony enters and shakes hands with each of the conspirators the blood is spread to his hands, and he marks himself out as someone who is also guilty of crimes against Caesar (he profits enormously from Caesar's death by becoming a ruler of Rome). The point of the scene may not be to realistically depict the death of Caesar, but to illustrate the event and dramatically draw attention to the guilt of those involved, who are left quite literally with blood on their hands.

Getting the blood onto the actors was one issue, but washing it off before the next scene presented another challenge. An example of how this was managed onstage comes from the play *Look About You*, performed by the Lord Admiral's Men in 1599. In the play the trickster Skinke goes through a series of transformations to evade capture. In one scene, he enters the stage disguised as Prince John, wearing the Prince's cloak and sword. This disguise is rapidly discarded, then Skinke changes clothes with a servant and applies blood to his face from a convenient saucer. He has only a few lines to achieve all of this before the 'real' Prince John enters. The Prince takes Skinke for a beaten and abused servant and lets him go. Before he leaves he tells the audience he will swim the Thames to Blackheath, hinting that he is going to wash off the blood; 250 lines later Skinke reappears without the blood and disguised as a hermit. Blood was clearly easier to apply than to remove, and the actor is given a considerable amount of time to clean up and don a new disguise.

In *Julius Caesar* Shakespeare allows much less time for the actors to clean up. Brutus has a 43-line opportunity to wash the blood from his hands and arms while Antony mourns over Caesar's body and discusses plans with his servant. Antony then has 34 lines to do the same while Brutus addresses the crowd in Caesar's funeral oration. The pace of the play isn't held up but there is still just enough time for the actors to wash off the blood – an example of Shakespeare's mastery of stagecraft.

Whatever the expectations and limitations of special effects in Elizabethan theatre, they didn't deter the Bard from including an extraordinary variety of different deaths for his characters.

CHAPTER THREE

Will You Be Cured of Your Infirmity?

> I know when one is dead, and when one lives.
>
> *King Lear*, Act 5, Scene 3

Shakespeare's plays are full of characters dying, both onstage and offstage. There are discussions about death too, and threats to kill.* Loss of life in the tragedies and histories is expected, but death lurks in the comedies too, where plays that are usually associated with laughter and romance also have characters who are threatened with execution (*Comedy of Errors*, *Measure for Measure*) or appear onstage in mourning (*Twelfth Night*). In *Love's*

* Of all the plays, only *The Merry Wives of Windsor* is completely free of death and any mention of it (although even here, Dr Caius does threaten to cut a man's throat and Mistress Ford jokes about someone being bitten to death by fleas). If you count Shakespeare's sonnets as a single work, only one of his poems, 'A Lover's Complaint', contains no allusions to death.

Labour's Lost, a play full of frivolity and witty word play, the announcement of a death brings the comedy to a juddering halt.

Advances in medicine have extended our average stay on earth considerably when compared to those living in Shakespeare's day, but death is still inescapable. No matter how much we may try to prolong or cheat the inevitable, 'All that lives must die, / Passing through nature to eternity' (*Hamlet*). The manner of our death, however, is likely to be very different from the experiences of our predecessors in the sixteenth century.

Today our final moments will most probably be spent in a hospital or care home where medically trained staff are close at hand. Sixteenth-century London had only three hospitals: St Bartholomew's, St Thomas's and Bethlem hospital for the insane. These institutions were strictly for the poor and few of those who entered expected to leave. Pistol's beloved Doll is the only one of Shakespeare's characters to go to a ''spital', where we are told she died of the 'Malady of France' (syphilis).* Her death comes in the same play, *Henry V*, as the demise of Shakespeare's great comic creation, Falstaff, but his death is the more typical experience of the day as he dies in a domestic setting surrounded by friends.

In the sixteenth century, and for some considerable time afterwards, births and deaths usually occurred in the home surrounded by friends, family and local gossips, rather than by the professional medical support expected today. Only the most shameful of illnesses such as syphilis,

* Some versions of the text have 'Nell' (Pistol's wife) instead of 'Doll'. See also Chapter 7.

or contagious diseases such as plague, kept visitors away. Few of those living during this time could have escaped witnessing death at close quarters.

Death itself was considered natural, with the exception of murder, suicide and witchcraft. But dying suddenly, alone or in disreputable circumstances was considered a bad death. This is not to say that the Elizabethans were complacent about dying and did nothing to avoid it. Medical treatments of various kinds were administered, wounds were treated by surgeons and attempts were made to revive the apparently deceased, as is shown in Shakespeare's plays.

★ ★ ★

The complexity of the human body means there are many points of vulnerability that can potentially bring about our end. Most of the time our bodies carry on doing their complex essential tasks without us really paying them much attention. But we are quick to notice when those processes aren't working properly, and seek medical help to make us feel better and prevent things becoming serious, or even fatal. The Elizabethans were no different in being preoccupied with their health and seeking medical help. The form of help they received, however, was very different.

There were many options for consultations open to patients in Shakespeare's day, but choice was governed by a patient's ability to pay rather than medical expertise. Options ranged from the highly regarded, university educated and very expensive physicians, to the much cheaper choice of the nearest wise woman. However, the size of the fee and education of the practitioner didn't improve the chances of recovery. The harm done by common forms of treatment generally outweighed any benefits linked with expertise or price.

In the sixteenth century and beyond, women were central to medical treatment and shouldered the bulk of the health-care burden. The range of female medical providers ran from friends and relatives offering basic nursing, to wise women who offered cures and treatments, and midwives who assisted at births. The skills of midwives were certainly highly regarded, and no pregnant woman would have considered having a male practitioner attend the birth other than under the most extreme circumstances. Despite this, midwives were certainly at the bottom of the professional medical hierarchy.

At the top were licensed physicians, called doctors because they had been educated at Oxford or Cambridge. Physicians learned from Latin texts, made diagnoses, treated internal ailments and prescribed medicines, but they did not cut into the body. Anything requiring a knife and involving bleeding was left to surgeons, found on the next rung down of the medical hierarchy.

Surgeons not only bled their patients, a common medical treatment of the day (see later); they also performed minor surgery such as removing stones, amputations, trepanning (drilling into the skull) and stitching up wounds. Barber-surgeons cut hair as well as flesh, hence the traditional sign outside barber shops of a red and white striped pole, signifying the blood and bandages of their profession. At a similar level of respectability were apothecaries who made and sold remedies, and may have carried out some unofficial diagnosing.

As well as medical people with accepted professional status, right at the bottom of the pile were all manner of unlicensed healers. Quackery was rife, but the lack of a licence did not necessarily mean the individual was a quack. Many cheap, and sometimes effective, treatments might be offered by women and men who had no official training, but vast experience.

All of these various types of medical practitioner are depicted, or at least mentioned, in Shakespeare's plays. Apothecaries appear on stage to sell poisons; surgeons are sent for to treat the wounded after altercations with swords; wise women are consulted; and doctors are depicted, both real and fictional. Shakespeare went further than all of his contemporaries in his portrayal of medicine and use of medical terms in his plays. There are hundreds of medical references in his work, direct and oblique, showing an understanding of health and anatomy far beyond that of any other playwright of his day.

One example of Shakespeare's in-depth medical knowledge was his apparent references to the theory of the circulation of the blood. This discovery is usually attributed to William Harvey and was first described by him in a lecture he gave in 1616 (although Harvey didn't publish his theory until 1628, after the playwright's death). However, the theory was certainly known before 1616 by a few medical men in mainland Europe and is hinted at in several Shakespeare plays, decades before it was accepted by the English medical establishment. Lines such as, 'You are my true and honourable wife; / As dear to me as are those ruddy drops / That visit my sad heart' (*Julius Caesar*) and 'The tide of blood in me / Hath proudly flow'd in vanity till now' (*Henry IV*, Part II) certainly suggest an appreciation of blood flowing, even if Shakespeare didn't explicitly state that blood flows in a continuous circuit.*

Where Shakespeare gained his knowledge has been speculated over for centuries. Some have suggested that he must have known William Harvey personally to gain insight into his theories. There is, however, no proof that they were acquainted. There were no compendium-style medical textbooks to study, but there were a lot of written treatises on specific ailments and medical theories that he could consult. Questioning medical practitioners directly might have been another route. Aside from personal experience of consulting medical practitioners for himself and his family, one doctor Shakespeare certainly did know was Dr John Hall, who married his daughter, Susanna, in 1607. The medical information contained within his plays certainly became more detailed after their wedding.

Shakespeare often poked fun at medical practices, the state of medical knowledge and the dire remedies that were doled out, but he held the doctors themselves in high esteem, with one notable exception. Dr Caius in *The Merry Wives of Windsor*, a pompous, self-important man much ridiculed by those around him, may be based

* Such detailed knowledge has often been cited as reason to suspect that William Shakespeare is not the true author of the plays and poems attributed to him. It has been argued that it would be very difficult for someone of his position and education to gain such knowledge. Another candidate for the authorship, the Earl of Oxford, certainly did have access to this knowledge, but he died in 1604 and Shakespeare continued to write plays for another 10 years.

on Dr Theodore de Mayerne, a prominent French physician who treated several French and English sovereigns. De Mayerne was president of the College of Physicians and apparently appeared very scholarly and sure of himself; perhaps he was a figure ripe for having a little fun made at his expense.*

Notable among Shakespeare's collection of doctors, quacks and apothecaries is the (fictitious) female practitioner, Helena, in *All's Well That Ends Well*, who successfully treats the King of France after all other male physicians have failed. Indeed, the doctors made him worse. Her medical knowledge is said to come from her physician father, Gerard of Narbon. Women were officially barred from studying medicine but knowledge was often shared through correspondence between practitioners and women. The character and the story of *All's Well That Ends Well* are not original to Shakespeare, who adapted the tale from Boccaccio's *Decameron*. Both versions illustrate how many women were respected for their medical knowledge, gained through informal methods of learning and sharing information, as well as personal experience of treating the sick.

Women are portrayed in all the usual medical roles that might be expected of them in Shakespeare's day. Though they don't appear on stage, wise women are mentioned in both *Twelfth Night* and *The Merry Wives of Windsor*. *Twelfth Night*, and several other plays, also includes one of the more curious methods used for

---

* Another real-life doctor Shakespeare depicted, though in a decidedly more favourable light than Dr Caius, was Dr Butt in *Henry VIII*, real-life physician to the King.

diagnosing disease and prognosticating over the fate of sixteenth-century patients – uroscopy. Flasks of a patient's urine would be collected and sent off to a doctor or wise woman for inspection, just as Falstaff does in Part II of *Henry IV*:

> Falstaff: Sirrah, you giant, what says the doctor to my water?
> Page: He said, sir, the water itself was a good healthy water, but, for the party that owed it, he might have more diseases than he knew for.

The upper echelons of the medical hierarchy turned their nose up at uroscopy but it continued to be popular. Examining the colour, clarity and odour of urine was supposed to offer clues to the state of the patient's health and temperament. Elaborate colour-wheels of shades of urine were produced to assist diagnoses. As Speed says in *Two Gentlemen of Verona*, 'these follies are within you and shine through you like the water in an urinal, that not an eye that sees you but is a physician to comment on your malady.'

In reality, only the most crude assumptions could be made by this type of examination and uroscopy certainly didn't deserve the trust many placed in it. Blood in the urine obviously indicates a serious problem with the kidneys, a sweet taste would be caused by diabetes and blue or dark urine can be a sign of porphyria or other diseases, but it is far from an exact science. Testing urine today can reveal vital information about a patient's health but this was simply not possible before the advent of modern analytical techniques.

Instead, the College of Physicians recommended measuring the pulse and judging fever by feeling the patient's forehead with the back of the hand, neither of which were terribly effective without stethoscopes and thermometers. All a physician really had at his disposal was his knowledge and observation through visual inspection, smell, taste and listening to the patient's complaints.

After diagnosis a treatment would be recommended to the patient and could differ greatly depending on which school of medical thought the doctor subscribed to. There were two theories of medicine battling to save lives in the sixteenth century. There was the traditional Galenic system of humours: good health was enjoyed by those who had the correct balance of four humours – black bile, yellow bile, phlegm and blood.* An imbalance of humours brought on illness and so physicians tried to restore the equilibrium through recommending special diets, blistering, sweating, purging (vomiting and excreting) and bleeding. The system of humours was more or less abandoned by the late seventeenth century, even though patients continued to be bled long after that. But traces of it can still be found in modern English; we still talk of good and bad humours to describe a person's temperament. Such phrases would have had a more literal significance in Shakespeare's day.

The alternative approach to health competing for attention in the sixteenth century was a new theory from Paracelsus (a Swiss physician almost exactly

---

* These were related to the four elements: earth, air, fire and water.

contemporaneous with Shakespeare), based on observations and a belief that everything was due to chemical processes (though his chemistry was very different from the modern science). He pioneered the use of synthetic chemicals and minerals as medicines.

Shakespeare was well aware of both theories and satirised the competition between the two in *All's Well That Ends Well*, 'Why, 'tis the rarest argument of wonder that hath shot out in our latter times.' But, with a few notable exceptions, such as laudanum,* there was little medical benefit in any of the treatments on offer at the time, and some of them threatened considerable harm to the patient. As Timon of Athens puts it to a thief, 'Trust not the physician; / His antidotes are poison, and he slays more than you rob.'

Helena's treatment for the King in *All's Well That Ends Well* claims to be completely harmless and promises to cure the King's fistula (an abscess in the chest) within two days. This benign remedy, whatever it might have been, and its rapid and complete success, is in stark contrast to the medical treatments available at the time. Regardless of whether the Galenic or Paracelsian system was used, treatment was for symptoms and not the underlying disease. There was little concept of disease as a distinct entity. For example, fever was seen as an illness in itself, rather than a symptom exhibited by many different diseases.

---

* Laudanum was invented by Paracelsus. It is opium mixed with alcohol. The combination is particularly effective in pain relief and became a staple of medical treatments for centuries.

People did not expect healers and medicine to cure them and they certainly didn't think that taking medicine would make them feel better. Most of the medicines that were prescribed would have made the patient feel very ill indeed. They would have been well advised to take Macbeth's advice and 'throw physic to the dogs'.

Even though it must have been evident that medicines rarely worked, they usually made the patient feel much worse, and people often died during treatment, the healer was rarely blamed for the death. Unsatisfactory results were explained away either because the malady was too severe for the treatment, or because the patient had failed to follow the often very detailed medical advice correctly. One surgeon, Tristram Lyde, ended up appearing in a Rochester court after he prescribed mercury treatments for several women suffering from syphilis who subsequently died. His defence was that the women were gravely ill and had failed to follow his instructions properly. The judge accepted the explanation and Lyde walked free.

Until germ theory was developed in the nineteenth century, medical practitioners were left guessing as to the cause of many diseases and therefore had little hope of treating them. Advances in medical treatment were also severely hindered by an almost complete lack of knowledge of physiology or pharmacology.

In spite of their best efforts, before the beginning of the eighteenth century, medical professionals had little impact on the population they treated. Death was rarely stopped or even slowed in its progress. Even identifying when death had occurred could be tricky, as we shall see.

★ ★ ★

In the overwhelming majority of cases death was easy to identify – vast personal experience would have made the process of death easily recognisable. Mostly, death in England before the sixteenth century was the concern of individuals, and their family and friends left behind to grieve. The state had little interest in the passing of its citizens. Things changed as the population grew and burial space, particularly in the capital, diminished. From 1538, parishes throughout England were required to record weddings, christenings and burials, but these records were kept locally without any wider coordination. However, burials became of acute interest during times of plague.

In the major outbreak of 1592, authorities in London started collating weekly tallies of burials within the parishes of the city, called the 'Bills of Mortality'. The practice continued until 1597 when the plague abated and was revived again in 1603 when the pestilence returned to the capital, and records were uninterrupted thereafter. In 1611, King James gave the task of producing annual mortality statistics to the Worshipful Company of Parish Clerks. In 1625 they were allowed a printing press so that the weekly bills could be published and the scale and statistics of death could be more widely known. However, the Bills of Mortality only focused on the 96 parishes within the city walls and 13 without.

As time went on more detail was added to the records, such as the age of the deceased and the cause of death. Determining the cause of death was the job of searchers, women employed by the parish to visit a home when a death was announced to inspect the corpse and collect information to report back to the parish clerks. According to their reports from the early seventeenth century, one could die of such diverse ailments as smallpox, plague, old age, grief, lunacy and 'teeth'.* Searchers received no medical training and many historians have subsequently decried their level of expertise and therefore mistrusted the information produced in the Bills. But, as we have seen, women were highly regarded in terms of medical knowledge and besides, many of the sick would have already been visited by some kind of medical practitioner. The cause of death would very likely already be known by the deceased's family and the searchers were simply gathering information from them. Mostly their job was to seek out cases of plague and check for signs that someone might not have died of natural causes.

Even in their limited state the Bills of Mortality have proved to be a fascinating and useful resource for researchers ever since. Today, detailed government statistics are recorded and determination of death and its causes is the province of professionals. The level of training invested in doctors and pathologists shows how difficult and complex their job can sometimes be.

★ ★ ★

Despite death being a universal experience, with millions of witnessed examples and centuries of scientific study, it is surprisingly difficult to define exactly what it is. As the author Christine Quigley put it recently, 'the dead are most often just like us, minus life'. From a technical

* No, I have no idea either. Your guess is as good as mine.

point of view, death is the complete cessation of vital processes within the body, but many processes continue long after a person is generally accepted to be dead. Cells die at different rates; neurons (nerve cells) are the most vulnerable, dying within minutes, but cells such as fibroblasts (connective tissue cells) can survive for days. Death is a process, not a single event and, depending on the circumstances, this process can take a relatively long time. Waiting for all these functions to cease is not generally necessary for a person to be declared dead.

The two key organs for determining death are the heart and the brain. When the heart stops beating, oxygen-rich blood can't be pumped to the brain, which stops functioning when starved of oxygen. Conversely, if the brain, and importantly the brain stem, is damaged, the body no longer receives instructions to breathe, and the lack of oxygen entering the bloodstream soon causes the heart to stop. Either way the result is brain damage that current medical interventions cannot reverse.

Medical advances in the twentieth century mean that hearts can be restarted and breathing can be supported if the brain is damaged. The heart, as long as it continues to receive oxygenated blood, will continue to beat even if the brain is no longer functioning. The boundary between life and death has become blurred and so further medical definitions were needed to pinpoint when a person is actually dead. Nowadays many countries define death as the cessation of brain function, but in the past it was generally taken to be the point when the heart stopped.

The period between cessation of the heart activity and brain death is termed 'clinical death'. This is when cardio-pulmonary resuscitation (CPR) can be used in an attempt to restart the heart and save a life. Without intervention the period lasts mere minutes before the brain, starved of oxygen, is irreversibly damaged. Modern devices such as the electrocardiograph and electroencephalograph can be used to detect faint electrical signals and activity within these organs, showing signs of life, but of course these were not available in Shakespeare's day.

The critical point at which nothing further can be done for a patient and death is inevitable would have been very different in the sixteenth century. That isn't to say that recovery was impossible, but there were no reliable standard procedures such as CPR to try to save an individual from dying. This certainly didn't stop the Elizabethans from trying. Those who were wounded would be carried to the nearest surgeon, if they could afford it. If it was thought that a person might be unconscious, attempts were made to revive them by warming the body or forcing medicine down the victim's throat. These methods were worlds apart from CPR, blood transfusions, defibrillators and modern medicines that can intensify weak heartbeats or reverse overdoses, but sometimes they were successful.

Revival from apparent death was not a new idea, even in Shakespeare's day. Pliny the Elder, a Roman author and natural philosopher, wrote in his *Natural History*, a book Shakespeare possibly read, of several cases of those who were carried to their funeral only to revive. On one occasion the heat of the fire was so strong that the presumed deceased revived, only to be consumed in the intense flames.

In *Pericles*, there is an example of someone being given up for dead who is later revived, not by incantations or the heat of a funeral pyre, but by the skill and knowledge of a physician. In the play, Thaisa is travelling on board ship with her husband Pericles during a storm when she gives birth to their daughter. The nurse that attends her believes Thaisa has died during childbirth and Pericles accepts the terrible news without question.

The sailors, mindful of the superstitions linked with carrying a corpse on a voyage, wish to commit the body to the waves before the storm destroys them all. Pericles complies, and Thaisa's body is washed, dressed in her finest clothes and placed in a chest that has been carefully treated and sealed to keep the water out. According to Lord Cerimon, who later finds the chest washed up on the shores of Ephesus, 'They were too rough, / That threw her in the sea'.*

When the chest is opened, Thaisa's body is found inside, but without the most obvious and conclusive sign of death – decay. Decomposition begins rapidly after death and can be detected early on by the unmistakable foul smell of compounds such as cadaverine and putrescine. These chemicals, among others produced during the decay process, have a power and a pungency that is difficult to ignore. By contrast, Thaisa's body in the chest is said to be sweet-smelling.

---

* Cerimon may be another character Shakespeare based on a real-life physician, Edward Stanley, the Earl of Derby, an amateur physician who 'was famous for chirurgerie [surgery], bone-setting and hospitality'. Another real-life candidate is Lord Lumley, who founded the Surgery Lecture at the College of Physicians in 1582.

Surprised by the absence of the smell of decay from what is supposed to be a dead body, Cerimon examines her more closely and detects signs of life. He says he knows of cases of Egyptians who were revived after lying apparently dead for nine hours, and from his examination of the body, he is sure Thaisa has been in her entranced state for less than five hours. With the help of a warm fire and the contents of his medicine cabinet she is restored to life. The story is fantastical but there are at least some aspects that have scientific credibility.

There is nothing surprising about Thaisa's death – death in childbirth was common enough in Elizabethan times. Even her revival after appearing dead is possible, though improbable. One potential explanation is that the loss of blood during childbirth leads to ventricular fibrillation. But, when the chest containing her body smacks into the waves, the force of the impact gives the heart a jolt that starts it beating normally again. However, without a blood transfusion, Thaisa would be unlikely to survive for long.

Another possible explanation comes from the human response to extreme stress or physical trauma, which can cause the body to shut down to try to preserve life for as long as possible. Perhaps this is how Thaisa survives her time drifting in the sea. Cold water can also put a body into a kind of stasis, allowing someone to be revived later. At body temperatures below 32°C a patient can lose all brainstem reflexes, be bradycardic (have a very slow heart rate) and be unable to shiver, depending on the degree of hypothermia. This has led to the axiom that someone is only truly dead when they are warm and dead. The phenomenon is well documented but unlikely to apply to Thaisa, as she was pitched overboard into the warm waters of the Mediterranean.

There is the added problem of air, or lack of it, inside the chest. The space inside the average coffin is normally only sufficient for 20 minutes of breathable air. Thaisa's chest would have to have been very roomy or have a few air holes in the top to give her enough oxygen to make it to shore and be discovered.

However it happened, Thaisa survived long enough to be discovered and was lucky to be found by someone with the expertise to revive her. Cerimon's methods of restoring her to life also have some scientific credibility. The warm fire could revive her in the event that she is suffering from hypothermia. Also, Cerimon's medicine cabinet could potentially contain drugs that would speed up a sluggish heartbeat and intensify contractions of the heart muscle. Belladonna has been used in medicine for centuries. It was prescribed for all manner of conditions, though mostly to treat inflammation and to relieve pain. Atropine, extracted from belladonna and several other plants common to Europe, is used today in emergency medical treatments to improve contractions of the heart, but this is a relatively recent medical discovery. Shakespeare, and therefore Cerimon, was probably unaware of the potential benefits of using belladonna.

The play ends happily with Thaisa being reunited with her family after many trials and tribulations. These days her traumatic experiences would lead to lawyers getting involved and lawsuits issued against those who had falsely declared her dead and thrown her into the sea. But in the play no blame is attached to the nurse, or to the sailors. Everyone in the play accepts that Thaisa is dead. A particular case from the seventeenth century shows how credible the story would have been to Shakespeare's audiences. In 1645, Françoise d'Aubigné, the daughter of a French governor, was thought to have died while at sea. She was sewn into a sack and was about to be dropped overboard when a meow was heard coming from inside the sack. The girl's pet cat had crawled in before the sack was sealed, and when they re-opened it to release the moggy, they realised the child was still alive. It just shows how difficult it was to be sure.

⋆ ⋆ ⋆

The risks of childbirth were well enough known to Elizabethan audiences and Thaisa's death wouldn't have been surprising. But when there was no obvious cause, no signs of disease or injury, it was all the more difficult to establish whether someone was really dead.

In *The Winter's Tale* Hermione dies of grief (more of this in Chapter 10) and there are calls to try and revive her, 'if you can bring / Tincture or lustre in her lip, her eye, / Heat outwardly or breath within', but to no avail. There are four indications of death given in this short speech: pallor, lustreless eyes, the loss of animal heat and no sign of breathing. Though Hermione receives no medical attention, that isn't the end. Towards the end of the play, decades after her death, a statue of Hermione is brought to life and she is reunited with her husband and daughter. This is the realm of fantasy, far beyond modern ideas of emergency medicine, which try to restore life only to the very recently deceased.

Hermione's revival is brought about by incantations and charms, and is perhaps closer to Frankenstein-type experiments that reanimate lifeless matter. Her situation may never have been intended as anything but fanciful and was an excellent excuse for special effects and a bit of stage magic. Nevertheless, the signs of death given in *The Winter's Tale* are not unreasonable, but they can also be exhibited by people when they are alive. For example, bodies are sometimes pale, and a pale complexion can be due to a non-fatal illness. Dead bodies may not be cold, and living bodies may not be warm, depending on the circumstances.

The most obvious signs to look for were lack of movement, breathing and pulse. However, those still alive but in catatonia (a state of stupor or unresponsiveness) can have a rigid immobility that can be mistaken for rigor mortis, and conversely, rigor mortis does not always occur in dead bodies. Another possible explanation for lack of movement, coma, is often diagnosed when the brain is rushed with endorphins to deal with extreme pain thresholds. The comatose person is not capable of moving their limbs; some nerves may twitch and shallow breathing can continue, but they certainly look dead to the untrained medical eye. Shallow breathing can be difficult to detect and prolonged listening with a stethoscope over the trachea or lungs is necessary.

Stethoscopes only became available in the nineteenth century, but Shakespeare details several crude alternatives that would have been used in the sixteenth century. For example, King Lear, desperate to believe that his daughter Cordelia is still alive, asks for a looking-glass to see if her breath will fog the surface. He also holds a feather to her mouth and nose and believes he sees some faint current of air disturb the soft barbs, but tragically he is mistaken. Yet another method that might have been tried in Shakespeare's day was placing a bowl of water on the chest and watching for tiny disturbances on the surface.

If a pulse couldn't be detected by fingers pressing against the skin, there were other methods that might be used. For example, string could be used to tie off the ends of the fingers, and if there was still a pulse, though faint, the tissue beyond the string would swell or change colour. A more accurate, but unpleasant way of checking for a pulse was to cut into a vein and see if blood flowed, but care had to be taken to ensure the blood wasn't just flowing because of gravity.

It is also possible to carry out crude checks for basic brain function, without sophisticated technology, by testing certain reflexes. For example, shining a bright light in the eyes normally causes the automatic contraction of the pupil. The eyes can also be tested for 'oculocephalic reflex', or the 'doll's head reflex', by holding them open and moving the head rapidly from side to side. If the reflex is present the eyes move in the direction opposite to head movement, but if they remain in the midline it is a sign of brain death. These kinds of tests have been around for centuries, but modern techniques are capable of determining even tiny eye movements.

After death the pupils dilate as muscles relax, and the eyes soon lose their roundness and take on an appreciable flatness. The surfaces of the eyes lose their gloss and a dull film appears. These are the earliest signs of decomposition and this may be what Shakespeare is referring to when he mentions Hermione's lustreless eyes.

In Shakespeare's day, waiting for a corpse to start showing signs of decay was pretty much the only failsafe option for knowing that a body was definitely dead and not just in a deep coma.* Given the difficulty of determining death back then, it is not so surprising that several of Shakespeare's characters are mistaken for dead and others convincingly fake it.

★ ★ ★

In *Cymbeline*, Imogen, disguised as a man named Fidele, is feeling under the weather and so swallows medicine she hopes will revive her. But, due to a complicated series of events (see Chapter 8), the medicine is actually a drug intended to give the appearance of death. Her unresponsive state is convincingly death-like and she is buried, but later in the play she is restored to health and reunited with her husband and family. It is a similar set-up to the most famous fake death in Shakespeare, that of Juliet in *Romeo and Juliet*, but with a decidedly happier ending.

Juliet's case is slightly different from others in Shakespeare's plays as she deliberately attempts to appear

---

* In the nineteenth century German fear of premature burial prompted the construction of waiting mortuaries. Strings were attached to bells, that were then tied to fingers, to signal when someone revived (no one did). In sixteenth-century England, mistakes were presumably rare enough, and waiting for decomposition was distressing, so bodies were buried before serious decomposition set in.

dead, rather than appearing dead by accident. The drug given to her by the Friar will not kill her but will make her seem dead for 42 hours. Judging by how often this was used as a plot device, and not just by Shakespeare, the idea that poisons could be tempered in some clever way to bring someone close to death but not kill them was a common one.

The story of *Romeo and Juliet* was not original to Shakespeare. As with the majority of his plays, he borrowed and adapted from existing work. There were many versions of the tale of the two star-crossed lovers from rival families circulating in the sixteenth century. Shakespeare's may not have even been the only version of the story being acted out on the Elizabethan stage, though it is the only theatrical version that has survived.

Some claimed the story to be true, or at least based on real events. Girolamo Dalla Corte dated the events to 1303 in his *History of Verona*, published between 1594 and 1596, and there certainly were two warring families in Verona at that time. But there is no evidence that the tale of the two lovers from opposing families is real. Each new version of the tale appeared with embellishments, extra characters and details added. The manner of the lovers' deaths also varied. For example, in one version Juliet, instead of stabbing herself after discovering Romeo's poisoned body, holds her breath until she expires, something that wouldn't actually be possible. The time of Juliet's unconsciousness also varies between 16 hours and two days. But the idea that she deliberately faked her own death is constant.

The details of the substance she used also varies. In some versions she is given a powder to dissolve in water, others have a liquid added to wine, or concoctions swallowed undiluted. Frustratingly, none of these versions gives a name to the substance, real or imagined, that could be used to do this. Without a name we can only use symptoms to guess at any substance that might have inspired such an idea.

Shakespeare's version of the well-known story is taken mostly from the poem by Arthur Brooke that contains many details not found in other accounts.* Even so, there are elements, such as the death of Paris at Romeo's hand outside the Capulet tomb, that are unique to Shakespeare.

In Shakespeare's play, it is the Friar who comes up with the plan and gives Juliet the vial of 'distilling liquor', describing the symptoms she can expect:

> all thy veins shall run
> A cold and drowsy humour, for no pulse
> Shall keep his native progress, but surcease:
> No warmth, no breath, shall testify thou livest;
> The roses in thy lips and cheeks shall fade
> To paly ashes, thy eyes' windows fall,
> Like death, when he shuts up the day of life;
> Each part, deprived of supple government,
> Shall, stiff and stark and cold, appear like death:
> And in this borrow'd likeness of shrunk death
> Thou shalt continue two and forty hours,
> And then awake as from a pleasant sleep.

From the symptoms, it sounds as if she is given a strong sedative. Today comas can be induced and maintained

---

* Brooke in turn based his version on a French translation of another version of the story written by the Italian Matteo Bandello.

using controlled doses of barbiturates but these were not available until the twentieth century. One sedative in use in Shakespeare's day, and for a long time before, was prepared from the roots of the mandragora plant. The plant was well known to Shakespeare, as he mentioned its sleep-inducing properties in both *Othello* and *Antony and Cleopatra*: 'Give me to drink mandragora / That I might sleep out this great gap of time / My Antony is away.'

Juliet herself talks about 'Shrieks like mandrakes torn out of the earth'. But she is referring to the myth that claimed the plants were the living link between plants and animals (underneath the green foliage its bifurcated root looks like a pair of legs), and that they screamed when they were uprooted.*

The root of the mandragora plant was used for millennia to induce unconsciousness; there are even references to it in the Bible. The active components in the plant include atropine, hyoscine and hyoscyamine, which all act in a similar way by blocking chemical signals to nerves in the central and parasympathetic nervous system. It could help with sedation, particularly when used with other drugs like opium, and it would cause the pupils to dilate, which also happens after death. But in other respects it would be a poor mimic of a death-like state. Juliet's heart would race and she was more likely to be flushed and hot than pale and cold.

Opium also sedates and slows breathing, but it causes pupils to contract, and the time of sedation is around

---

* The screams were thought to be so terrible that they could kill a man, or as Shakespeare put it in *Henry IV* Part II, 'Would curses kill, as doth the mandrake's groan'.

eight hours rather than the 42 hours in the play. Upping the dose does not prolong the period of unconsciousness. Instead, it intensifies the level of sedation, to the point where, in the case of an overdose, breathing stops.

Shakespeare's play, and other sources, suggests that normally fatal poisons could be moderated in some way so that they would stop short of killing someone, but this is simply not possible. Chemical reactions can be used to manipulate the structures of compounds extracted from plants, to make them less toxic and minimise side effects. But in the sixteenth century it wasn't even known what the active substances were within the plant, let alone how they might be isolated and modified.

There is, however, one compound that has been known to give a very convincing appearance of death: tetrodotoxin (TTX). There are many cases of humans ingesting TTX and being certified dead before later reviving. TTX blocks the transmission of nerve impulses along nerve fibres to muscles, including the diaphragm. If muscles do not receive nerve signals they will not contract, and therefore the poison can cause respiratory paralysis and death. There is a very narrow window where breathing, and heartbeat, can be slowed to almost undetectable rates, but not stopped completely. The brain may continue to function, and the individual may be awake and aware of what is going on, but cannot signal their distress. As the body slowly breaks down the toxin from the body, normal nerve function returns and a person can revive.

TTX is a heat-stable neurotoxin produced by bacteria. Some animals that come into contact with the bacteria, including the blue-ringed octopus, some species of salamander and the Californian newt, accumulate the toxin deliberately. These animals use it either as a defence mechanism or as a way of paralysing prey so they can be more easily caught and eaten. TTX poisoning in humans is most commonly associated with eating improperly prepared puffer fish. Puffer fish accumulate TTX in their liver, ovaries and skin. Preparation of puffer-fish sushi (*fugu*) requires great skill and years of training to prevent dangerous levels of TTX from being transferred to the flesh of the fish. Restaurants in Japan need special licences to sell *fugu* but, despite all the precautions, mistakes have been made, usually by fishermen who decide to eat some of their catch, or supermarkets that mis-identify fish being put on sale. The Japanese are therefore well practised in treating puffer-fish poisoning, which is done by artificially supporting breathing until the toxin is cleared from the body. Survival rates are now very good. Things would have been very different 400 years ago.

TTX may be a strong candidate for Juliet's death-defying poison but it falls down on one key point – puffer fish were not known in Europe until the time of Cook's voyages in the eighteenth century. One remote possibility is that trade with south-east Asia brought the knowledge, or rumour, of such substances into Europe much earlier and inspired stories of death-simulating potions.

* * *

Fake death or mistaken death was very much the exception rather than the rule in Shakespeare's plays.

In the vast majority of cases death was obvious. When Mistress Quickly sees Falstaff on his sickbed she says, 'I knew there was no other way' (*Henry V*). Several deaths in Shakespeare's works show he had a sophisticated appreciation of death as a process rather than a single event. In more than one character he describes the slow progress towards their death as the body shuts down, bit by bit.

Falstaff is a case in point. His final moments are described to the audience by Mistress Quickly in *Henry V*. He was delirious, his feet were cold and chill spread up his body as his heart failed to effectively circulate his blood. She adds, 'I saw him fumble with the sheets and play with flowers and smile upon his fingers' ends'. Plucking and picking at small objects is referred to technically as 'carphologia', and is a sign of extreme exhaustion or impending death.

It is a very mundane end to the exuberant life of the fat knight who featured in three of Shakespeare's plays.* Audiences are forewarned of his death at the end of *Henry IV* Part II, and they are told that Falstaff will die of a sweat, perhaps a reference to the sweating sickness, or even plague.

Famed for his love of women and sack,† Falstaff was hardly living a healthy life and threats to his wellbeing might have come from anything from venereal disease to alcoholism to diabetes, as well as all the usual hazards in

---

* Both parts of *Henry IV* and *The Merry Wives of Windsor*. Falstaff doesn't actually make an appearance in *Henry V*.

† Sack is an old term for white wine imported from Spain that was fortified with brandy.

Middle Ages England. But no specific cause of death is given as we would understand it today. Mistress Quickly says Falstaff was suffering from a 'burning quotidian tertian'. This is a very precise diagnosis, for the sixteenth century anyway, and describes a combination of fevers (or agues) that were considered particularly dangerous. These symptoms could have been due to the fever caused by the sweating sickness, or maybe malaria finally felled this great Shakespearean character (see Chapter 7). Another suggestion is that the delirium described in the play was caused by typhoid.

Whatever the cause, Falstaff's life of excess certainly wouldn't have helped his chances of survival. The play suggests that Henry V (Prince Hal in *Henry IV*) also contributed to the death by breaking Falstaff's heart when he so callously dropped his friendship with the knight the moment he ascended to the throne.

Another drawn-out death portrayed in considerable detail by Shakespeare is that of Henry IV. At least this time Shakespeare was depicting a real-life person and we have historical records to refer to in an attempt to get at the truth. But still, the exact cause of death is unknown and has been speculated about by scholars ever since. The only thing that is certain is that it was due to natural causes.

Henry IV suffered from several illnesses during his life, some of them psychosomatic. However, in 1408 the King suffered a mild stroke and his health deteriorated from that point. He suffered fainting fits and some kind of heart complaint, and was occasionally incapacitated to the point that he could not walk. He was certainly under considerable stress; the insecurity of his position on the throne and constant criticism from those who considered him an usurper can't have been helpful to his mental health. Some have asserted he was struck down with leprosy because of eruptions on his skin. It was said he became so disfigured that no one could bear to look at him. The French believed his toes and fingers had fallen off, another sign of leprosy. At least these reports have some semblance of credibility. Other accounts of his illness are more ridiculous – the Scots said that he had shrunk to the size of a child.

However, his death mask, and the examination of his well-preserved face during the exhumation of his body in 1831, suggests that accounts of his disfigurement were exaggerated. One modern medical opinion is that it could have been tubercular gangrene (in rare cases tuberculosis can lead to gangrene), combined with erysipelas (a skin infection, typically with a rash), which produces a burning sensation. This would at least explain accounts of him crying out in agony saying he was burning.

His last illness began when he was taken sick very suddenly on 20 March 1413, when praying at St Edward's shrine in Westminster, and was carried to the nearby Jerusalem chamber. He may have suffered another stroke as it states it took him a little while to recover his speech. He was placed on a makeshift bed near the fire to keep him warm but he complained his arms and legs were cold – it was clear to everyone that he was dying.*

Shakespeare's account of Henry's final illness from *Henry IV* Part II seems to have been fairly accurate. As

---

* Henry IV later died on the same day he was taken ill.

shown in the play, the King had periods of unconsciousness and during one of these, his son entered the room to take his crown, believing him to be dead. In his dying moments he complains of loss of sight, and wasted lungs – his body is shutting down as the heart fails to pump oxygenated blood effectively. It is similar to yet another Shakespearean depiction of a real-life death, Katherine of Aragon in *Henry VIII.* She says in her final moments, 'Mine eyes grow dim' – parts of her brain, deprived of oxygen, are starting to fail and death is close.

The ultimate cause of all deaths is the failure of oxygen to reach vital structures within the body. As Dr Milton Helpern, Chief Medical Examiner of New York City, succinctly put it, 'Death may be due to a wide variety of diseases and disorders, but in every case the underlying physiological cause is a breakdown in the body's oxygen cycle.'

Elizabethans, however, had no concept of oxygen, because the element wasn't discovered until 200 years after Shakespeare was writing. But the idea of cessation of breathing as an important indicator in determining death was well established. The necessity and life-giving properties of blood were also well known. The idea of the 'life of the blood' has a history going further back than Hippocrates, to the Old Testament, where it is stated that 'the life of the flesh is in the blood'. However, the details of how the blood carried out its life-giving function were anything but clear.

In the moments around death, gravity slowly pulls the blood, and to a lesser extent blood plasma, to the lowest point in the body, 'descended to the labouring heart' as Shakespeare put it. In the intervening period the skin can appear mottled or patchy in colour due to uneven vasodilation. The loss of blood pressure leaves skin flaccid and pale, more eloquently put in *Henry VI* Part II as 'meagre, pale and bloodless'.

Without fresh supplies of oxygen, cells begin to die. The last few moments of life are often accompanied by a short series of heaving gasps, a final desperate attempt to get hold of oxygen. On rare occasions there may be a laryngeal spasm, which causes the 'death rattle', brief agonal convulsions, and the chest and shoulders may heave.

All death may come down to disruption in the oxygen cycle, but there are a huge number of ways it can happen. Shakespeare explored many of these different causes of death in his work. Some are true to life and others are created from his imagination, but all of them are interesting in their own way.

CHAPTER FOUR

Off With His Head!

The place of death and sorry execution

Comedy of Errors, Act 5, Scene 1

Public execution was a common event in Shakespeare's day. Convicted criminals were beheaded, hanged, burned, boiled and squeezed to death, and all in front of an audience. Scaffolds were sited at Newgate, Tyburn and other parts of London, and if you didn't witness the event itself, grim reminders could be seen as you went about your daily business on the streets of the capital. Severed heads looked down on pedestrians crossing London Bridge; the bodies of gibbeted murderers swung in the breeze. Tourists sometimes made the trip out to Wapping, where those convicted of piracy were hanged at the low-water mark and left there until they had been washed three times by the tide. It is no surprise that executions in all their grisly variety get a mention in many Elizabethan and Jacobean dramas.

★ ★ ★

Dozens of characters are sent to their deaths by execution in Shakespeare's plays, almost all of them in the histories. Traitors, thieves and witches all receive the ultimate punishment for their crimes. But the death penalty was such a common occurrence in Elizabethan and Jacobean England that Shakespeare even includes it in some of the plays normally considered comedies. In the *Comedy of Errors*, Aegeon of Syracuse travels to Ephesus in search of his long-lost sons, but a recent ban imposed on Syracusians entering the city leads to his arrest. Aegeon has broken the law and the punishment is death: 'if any Syracusian born / Come to the bay of Ephesus, he dies'. The sad tale of his search for his displaced family gains him some leniency and he is given one day to find someone who will lend him money to pay a hefty fine, but if he does not pay he will be executed.*

Death sentences for what may seem relatively minor crimes are not confined to the *Comedy of Errors*. Another play, *Measure for Measure*, also listed as a comedy in the First Folio, has become known as a 'problem play', not least because three men are condemned to death for sex outside of marriage, but only two are saved: 'Is any woman wrong'd by this lewd fellow, / As I have heard him swear himself there's one / Whom he begot with child, let her appear, / And he shall marry her: the

* Spoiler: it all turns out all right in the end.

nuptial finish'd, / Let him be whipt and hang'd.' The fact that the women in *Measure for Measure*, though equally guilty, are shamed and forced to marry (some more willingly than others), but not hanged, shows how unevenly the law could be applied in Shakespeare's day. This play, and several others, show the Bard's preoccupation with the process of law and how justice was administered, not just the final sentencing.

Some of the crimes may seem exaggerated and melodramatic to modern audiences, but they were not so far from reality in Shakespeare's day. Some would not be considered crimes at all today, and the punishments were certainly very different. Corporal (physical) and capital (death) punishments are no longer a feature of the British justice system, but 400 years ago both would have been a common sight. The playwright would have been familiar with corporal punishment even before he arrived in London: Stratford had whipping posts, pillories and stocks. His move to the metropolis, however, would have introduced him to a far greater number and severity of punishments.

By including executions in their plays Shakespeare and his contemporaries may have been simply reflecting the reality of life and death around them, but few playwrights were brave enough to depict the death itself onstage. Guilty parties are generally sentenced, then led offstage to their execution; then brief reports of their death might be given after the event, or severed heads are brought out and displayed to the audience. Staging such horrors would have been dangerous for the actors and, even with the best special effects, would not live up to the reality of capital punishment that would have been so familiar to the audience. There was also the possibility that seeing a popular figure executed onstage could provoke the already rowdy audience into more dangerous behaviour. Keeping the moment of death out of sight might have been a simple precaution for crowd control.

One notable exception is a play by Thomas Kyd, *The Spanish Tragedy*, which includes an onstage hanging. The play is credited as the original revenge tragedy and was hugely influential.* It was also enormously popular with audiences and was frequently staged, leading to much modern-day speculation as to how the hanging could be achieved without killing off a string of Elizabethan actors.

Shakespeare doesn't quite match Kyd's brutality. He has no executions carried out onstage, and any that are seen in modern productions are there because of an artistic decision, not because they are included in the text of the play. And, unlike other deaths in his plays, there is very little description or discussion of executions. This probably isn't due to any squeamishness on his part, but because his audiences didn't need the process described to them; they already knew what executions looked like.

The death penalty was not a straightforward sentence in the sixteenth century and there were many variations in methods of execution. The manner of state-sanctioned death was chosen based on the crime as well as the social

* Many playwrights made use of the new theatrical genre of revenge tragedy, including Shakespeare: *The Spanish Tragedy* may have been one of the sources for *Hamlet*.

status of the condemned. Shakespeare's often simple, even bland statements about capital punishment disguise an elaborate system of state execution that modern audiences are often unaware of, and this chapter fills in the background to something that Shakespeare and his audiences almost took for granted.

★ ★ ★

Punishments for crimes in sixteenth- and seventeenth-century England were severe and usually carried out in public, as it was thought it would deter others from committing the same crimes. The severity of the punishment was meant to match that of the crime. However, there were many discrepancies and loopholes that could see vicious criminals escaping with relatively minor sentences and the innocent suffering miserable deaths in prison.

Crimes in Shakespeare's era could be broadly classified into two types: misdemeanours and felonies. For misdemeanours, the punishment could be imprisonment, flogging, a fine, forfeiture of goods, or a combination of all four. Felonies were more serious, and therefore deserving of more severe punishment. The consequences of being convicted of a felony were, in those days, the loss of all possessions to the Crown, and death.

Crimes that were considered felonies ranged from treason – the most despicable of all – to theft of anything worth more than a shilling (twelve old pence).* The list

* One shilling was the average daily wage of actors and skilled artisans.

of possible felonies was a long one and consequently execution was a common occurrence. In Shakespeare's lifetime upwards of 1,000 hangings took place every year in England and Wales. All surviving evidence suggests that the Elizabethan and Jacobean eras had significantly higher levels of executions than later periods.

The number of executions may have been terrifyingly high but around three-quarters of those charged with a felony escaped the death penalty. One explanation for this apparent anomaly is that there was no close scrutiny of cases before they got to the courts and many would have been dismissed at this stage. There were also no guidelines for sentencing, as there are today, and judges had considerable freedom over the fate of the felons convicted in their courts.

Another important part of justice was mercy, a concept explored in detail in *The Merchant of Venice*. This play, which could almost be considered a courtroom drama, has Shylock loan Bassanio a large sum of money on the condition that the debt will be repaid in full in three months' time. If he can't pay, Antonio, who has stood as Bassanio's guarantor, must forfeit a pound of his flesh. When he fails to come up with the money Shylock takes Antonio to court to argue his case. The judge finds the agreement legal and there appears to be no way to prevent Shylock from taking his dues, even though Antonio is expected to die in the process. Portia, disguised as a lawyer, intervenes and advocates for mercy:

> The quality of mercy is not strained.
> It droppeth as the gentle rain from heaven
> Upon the place beneath. It is twice blest:

> It blesseth him that gives and him that takes.
> 'Tis mightiest in the mightiest; it becomes
> The thronèd monarch better than his crown.
> His scepter shows the force of temporal power,
> The attribute to awe and majesty
> Wherein doth sit the dread and fear of kings;
> But mercy is above this sceptered sway.
> It is enthronèd in the hearts of kings;
> It is an attribute to God Himself;
> And earthly power doth then show likest God's
> When mercy seasons justice.

Portia's eloquent pleas fall on deaf ears, Shylock is determined to get his way, and she must resort to an obscure legal clause to save Antonio. Despite her fine words, Portia shows precious little mercy when it comes to her treatment of Shylock and he in turn faces the death penalty unless he converts to Christianity.

In court other factors would also be taken into account. A pardon might be given if there were mitigating circumstances, such as if a death during a fight was thought to be as a result of misadventure or committed in self-defence. At the end of *Romeo and Juliet* many people are dead because of the actions of others. Prince Escalus declares that 'Some shall be pardon'd, and some punished'. As the dispenser of justice in the play it is his decision to make, but he doesn't elaborate on who will benefit from his power to pardon.*

Even if a felon was found guilty and received no pardon, there were still ways of escaping the death

---

* Speculation over who in the play deserves to be pardoned has been a gift for exam writers.

penalty. The accused could claim 'benefit of clergy' (see Chapter 1), a potential legal loophole that had arisen as one of the consequences of the conflicts between church and state in the eleventh to thirteenth centuries. The clergy obtained the right to be tried for certain types of felony in their own ecclesiastical courts, which did not have the death sentence, rather than the royal courts.

To prove himself a clergyman, all a man had to do was take a psalter that was handed to him and, in the presence of the bishop's representative, stumble his way through the 'neck verse' – so called because it could save someone's neck rather than in reference to the contents of the passage – proving he could read.* The bishop's representative would then acknowledge the claim. The standard of literacy demanded was not high and it offered a degree of compassion, given the harsh alternatives. The felon would still lose all his property, and might be jailed for up to a year, but at least he escaped with his life. He was also branded on the fleshy part of the thumb to prevent him from enjoying the same privilege twice.

The assumption that only those educated in the Church could read had been false for centuries, and it was grossly unfair that a man's sentence depended on his ability to read.† Shakespeare made a point of satirising the situation in *Henry VI* Part II. The fourth act of the

---

* Usually they were asked to read, in Latin, the first few lines of Psalm 51, which asks for forgiveness for past sins.

† The ridiculous nature of the law was further highlighted by the fact that in 1624 benefit of clergy was extended to women, even though the idea of a female cleric would have been laughable at the time.

play centres around a rebellion against King Henry's government led by Jack Cade.

Cade and his followers marched on London and fought a pitched battle on London Bridge. They successfully entered the City and Cade declared himself mayor. He then set up tribunals to determine the guilt or innocence of those accused of corruption. Although Shakespeare is largely faithful to the history, he borrowed some details from another earlier rebellion – the Peasants' Revolt* – to highlight the excesses of the rebels and how they threatened to overturn the everyday order of things. According to Raphael Holinshed's *Chronicles of England, Scotland and Ireland*, one of Shakespeare's main sources for his history plays, in the earlier revolt 'it was dangerous among them to be known for one that was learned, and more dangerous, if any men were found with a pen and inkhorn at his side: for such seldom or never escaped from them with his life.'

Shakespeare condenses this into the treatment of one of his characters and advances it 170-odd years to be part of Cade's unofficial tribunals. The man in question has been found with a book in his pocket – evidence that he can read. The clerk admits, 'I have been so well brought up that I can write my name.' Cade sees being literate as evidence that he must be 'a villain and a traitor', and issues orders to 'hang him with his pen and inkhorn about his neck'.

Some crimes were not clergyable, particularly the more serious crimes such as treason, murder with malice

* This popular uprising of 1381 was against the high taxation enforced by Richard II's government.

aforethought and rape. Over the years more crimes were added to the list but the number of non-clergyable crimes was still small. In the sixteenth century only sodomy, bestiary, witchcraft, picking pockets and horse-theft were non-clergyable.

Another way to avoid the death penalty was 'benefit of the womb'. Pregnant women would have the day of execution delayed until after the birth of their child. Nearly half of all women convicted of a felony claimed they were pregnant, 38 per cent of them successfully. It is possible that after the birth, when the time came to be hanged, the woman was given a full pardon and escaped death. In Shakespeare's *Henry VI* Part I, Joan of Arc claims to be pregnant when she is sentenced to be burned for witchcraft by the English and her execution is consequently delayed. When no baby appears after nine months the execution is rescheduled.

Benefit of the womb is an important plot device in *The Winter's Tale*. Hermione, imprisoned by her husband who thinks she is pregnant by another man, delivers her baby girl in jail. Paulina offers to take the child, 'If she dares trust me with her little babe, / I'll show't the king and undertake to be / Her advocate to the loud'st' in an effort to get Hermione released. But the jailer refuses to let the child leave the prison. Paulina protests, 'This child was prisoner to the womb and / By law and process of great nature thence / Freed and enfranchised' – the baby was innocent of the crime committed by the mother and therefore did not deserve the same punishment.*

* Spoiler: the baby girl is banished but grows up, falls in love, is reunited with her family and everyone lives happily ever after.

But if the crime was exempt from benefit of clergy or the womb, and there were no mitigating circumstances, then death was the usual punishment. The sentence would have been read out by the judge, 'Though shalt first return to the place from whence thou camest [prison], from thence thou shalt go to the place of execution, there thou shalt hang till thou be dead …' There was little time to try to gain a reprieve, although this was often granted. The ability to make such an appeal depended heavily on how wealthy the felon was and how much influence they wielded. A pauper stood little chance. Shakespeare's observations in *Measure for Measure*, where one convict has escaped the death sentence for nine years because 'His friends still wrought reprieves for him', are entirely plausible.

⋆ ⋆ ⋆

The process of law could also be very unpleasant, even life-threatening. Someone accused of a crime in the sixteenth and seventeenth centuries was carted off to prison to await trial and, back then, courts did not sit continuously. Quarter sessions, so called because they sat four times a year, settled the less serious infringements of the law. Cases of serious crime, such as murder and rape, were referred to the Assizes, courts that travelled around the country on a circuit. The accused would have to wait until the court was in session and as a result could sometimes spend months in prison before their case was even heard.

However, once justice arrived, it operated quickly. Trials were short and judgments enacted promptly.

The sight of Shakespeare's characters being removed from the stage to go to their execution immediately after sentencing is not so far from the reality of Elizabethan justice. If the accused were found not guilty they would be released immediately. Those that received a pardon, however, would be returned to prison until that pardon was issued, sometimes several months later.

Unlike today, imprisonment itself was not a form of punishment for serious crime. Those convicted of misdemeanours and those in debt could find themselves imprisoned until their sentence had been completed or their debt paid. But given the general conditions of prisons at the time, it was anything but a light sentence. The characters imprisoned in Shakespeare's plays are under no illusions as to the seriousness of their situation. In *Twelfth Night*, Malvolio, tricked into appearing mad, is locked up as revenge for his strict and overbearing ways. But the joke goes too far. He is kept in darkness and 'notoriously abused'. In *The Two Noble Kinsmen*, Arcite and Palamon are on the losing side after a battle. Captured and imprisoned, they despair at the loss of their liberty. They will never see their friends again or enjoy the comforts of their former life. Outside it might be summer but in prison 'dead cold winter must inhabit here still'. It is clearly no picnic but Arcite and Palamon's high social status means they are actually very well treated in prison – they eat well and have clean and comfortable accommodation (at least by Elizabethan standards). Not everyone was so lucky.

Elizabethan prisons were far from conducive to preserving the lives of their inmates. Conditions were appalling: crowded and filthy. The buildings themselves were often in a poor state of repair, sometimes so bad that escape was a realistic, and no doubt desirable, option. There was no exercise available for prisoners and only the most basic food rations were supplied. Meals could be supplemented by friends and relatives when they visited the inmates, and wealthier prisoners could pay extra to purchase food through the prison bars from sellers who congregated outside the walls. Even in times of plenty, those without means went hungry, and in times of dearth, very hungry.

Malnutrition, unsanitary living conditions and lack of exercise could severely debilitate a prisoner, leaving them particularly vulnerable to infections. The chances of dying of 'jail fever', a form of dysentery, were high. It was so virulent at times that in one famous incident in 1577, when sick prisoners were brought for trial in Oxford, they infected both jurors and judges, several of whom died. Surviving reports from the King's Bench show that between 1558 and 1625, a total of 1,292 prisoners died in the jails of the Home Counties alone. Some of these may not have been convicted of anything and were merely waiting for their opportunity to defend themselves in court, or for a pardon to be issued.

In *Measure for Measure* Ragozine, 'a most notorious pirate', suffers the fate of many inmates and dies in prison 'of a cruel fever'. His death is unremarkable for the time and scarcely commented on. What excites more interest is his facial similarity with Claudio, the hero of the play. Claudio has been sentenced to death, unjustly in the eyes of most people, and to save him from imminent execution Ragozine's head is cut off and sent to the authorities so they will believe Claudio's beheading has taken place. It's not clear at what stage of the judicial process Ragozine dies. But, if he really was a pirate as notorious as claimed, it was unlikely he would have lived long. Whether justice was better served by his death in prison or hanging at the end of rope at low-water mark is debatable.

★ ★ ★

The swiftness of executions might seem surprising as there was a lot to organise within such a short time frame. But because it was a relatively regular event, particularly in the capital, much of the infrastructure for execution was already in place. In more rural areas, where executions were less common, things might be delayed while everything was arranged. In 1655 the execution of one Captain Hunt was delayed while a scaffold was built and an axe procured of the correct length (11 inches was specified), by which time the prisoner had escaped.

But with regular repetition a pattern of events was soon established; the process of conviction, sentencing and execution followed a general, well-established format. Until as late as 1868 the majority of executions were conducted in public. They took place on high platforms in front of big crowds with sermons, speeches and a dramatic finale – it was very theatrical. In the normal run of events thousands of 'sorrowful spectators' gathered at the site of execution and it would not be unusual for individuals in the crowd to see people they knew on the scaffold. It was a socially diverse group that attended these events, just like the theatre. Noblemen might be in the crowd along with apprentices and pickpockets, though there may have been some slight segregation based on an individual's ability to pay for a better viewpoint.

When the crowd had gathered, there was a procession from the prison where the convicted had been held to the place of execution. This was followed by a sermon, read by a priest, and immediately after, a speech was often made by the prisoner, just moments before they were launched into eternity. The condemned was expected to make a 'good end', meaning to show courage at the gallows, and appear penitent and contrite. Given the extraordinary stress of the situation it is impressive that the majority of condemned felons did just that. When many might have been expected to take the opportunity to hurl abuse at those about to take their life and the injustice of the world, they often admitted to their crimes, along with past sinfulness, and exhorted others to learn from their example. The speech was usually credited with being absolutely truthful. These individuals were shortly to meet their maker, who would pass the ultimate judgment on them. This was the time for complete honesty.

In Shakespeare's *Henry VIII*, the Duke of Buckingham is found guilty of treason and sent to be beheaded. He makes the traditional impassioned speech, still proclaiming his innocence but full of forgiveness of his accusers, before he is led offstage to his fate. Buckingham's speech may seem lengthy (58 lines in two sections), and an excuse for the Bard to show off his eloquence, but this was not unusual. One observer complained that a felon had kept him, and everyone else, standing in the rain for half an hour while slowly talking through a lifetime of crime.

After beheadings the head would have been held up and shown to the crowd and, particularly if it was the execution of a popular or revered figure, handkerchiefs were dipped in the spilled blood as a keepsake.* When a prisoner was hanged the body would be left suspended for up to an hour to ensure death. Family members or friends would pull on the feet of their loved ones in an effort to shorten their suffering by hastening death. Signs of a swollen, blackened face and hands were looked for to be certain of death before the executed was cut down. Accounts of the death would be written in broadsheets and sung about in ballads for the benefit of those who could not be there to witness it in person.

⋆ ⋆ ⋆

While the punishment for any felony was always death, different crimes demanded different deaths. Up until the mid-sixteenth century punishments were carefully chosen to give a foretaste of what awaited guilty individuals in hell. By Shakespeare's day, boiling poisoners in cauldrons of water or lead and drowning witches was slowly falling out of favour, but there was still plenty of gruesome variety available. In general, the more severe the crime, the more the convicted could expect to suffer. But it didn't always work out that way.

The most despicable crimes were those committed against the king or queen's person. Plotting the overthrow or death of the sovereign was high treason, punishable by

---

* The blood of executed criminals was also thought to have curative properties.

the most extreme form of execution available at the time. In England this was hanging, drawing and quartering, a form of punishment that had been introduced in the thirteenth century and was only finally abandoned in 1817.

First, the convicted was hanged by the neck, but cut down before they were dead and were still able to watch as their entrails and heart were drawn out of their abdomen and burnt on a fire in front of them. The shock of pain and blood loss most likely killed them during disembowelling, or at least left them unconscious. Any life still clinging to the body would have been snuffed out shortly after the heart was removed. The head was then cut off and the body cut into quarters – literally hacked to pieces. The executioners, often butchers by profession, and those in the crowd who got too close, were left splattered in the guilty man's blood. The head and other sections of the body were stuck on spikes for public display to demonstrate the power of the Crown and act as a warning to anyone else thinking of plotting against their sovereign. Sometimes, if the treason was not considered so great, the condemned person was left to hang until they were dead before the rest of the process was carried out – a considerable mercy.

Only two of Shakespeare's characters suffer this gruesome form of execution.* The first is in *Henry IV* Part I. In the play the King's forces are facing the opposing army of Hotspur. In an effort to avoid battle and bring peace, Henry sends 'grace, / Pardon and terms

---

* Jack Cade from *Henry VI* Part II died during his arrest but his body was later cut into quarters for public display.

of love' via his messengers Sir Richard Vernon and Thomas Percy, the first Earl of Worcester. But Vernon and Worcester deliberately withhold the information, which leads to a battle and results in many casualties on both sides. Henry is victorious and Hotspur is killed. After the battle Worcester and Vernon are captured and their actions deemed traitorous. They acted against their King and the consequence was thousands of unnecessary deaths. Both are found guilty of treason and both are sentenced to be executed – 'Bear Worcester to the death and Vernon too'. However, Shakespeare makes no mention of how they are to meet their deaths.

There was no need to go into details; contemporary audiences knew perfectly well what happened to traitors. Audiences would also have understood the different social ranks of the two condemned men. Vernon is from a well-to-do family but Worcester is a nobleman. The play is based on real historical events, so it is easy enough to find out what would have happened to the characters once they left the stage. The real-life Sir Richard Vernon was hanged, drawn and quartered. But because of his noble birth, the real-life Earl of Worcester was beheaded – a much swifter end.

The other example of the ultimate punishment for the ultimate crime comes from *Henry VI* Part II. Eleanor, Duchess of Gloucester and wife to Duke Humphrey the Lord Protector, wants to know when the King will die so that her husband will take the Crown and she will become Queen. To find out the answers she employs a witch and a necromancer and persuades several of her staff to summon a spirit to tell her what the future holds. Those involved are Roger Bolingbroke, Eleanor's personal clerk, John Hume, Eleanor's personal chaplain, Thomas Southwell, another man of the cloth and Margaret Jourdain, also known as the Witch of Eye.* Eleanor, Bolingbroke, Hume and Southwell are all indicted for sorcery, felony and treason. Jourdain is tried as a witch.

The real-life plot took place in 1441 and Shakespeare followed the history fairly closely. The playwright compresses a series of trials into one moment where all the accused are found guilty and sentenced:

> You four, from hence to prison back again;
> From thence unto the place of execution:
> The witch in Smithfield shall be burn'd to ashes,
> And you three shall be strangled on the gallows.
> You, madam, for you are more nobly born,
> Despoiled of your honour in your life,
> Shall, after three days' open penance done,
> Live in your country here in banishment,
> With Sir John Stanley, in the Isle of Man.

Eleanor gets off lightly, as she did in real life due to her high social status, with penance and perpetual imprisonment. Hume, contrary to what Shakespeare says, was pardoned as he had really only played a supporting role in events. Southwell died in the Tower before his sentence could be carried out. But Bolingbroke, considered to have played a more significant role, was hanged, drawn and quartered and his head displayed on

---

* There are several spellings of Margaret Jourdain's name in the historical record including Margery Jourdemayne and Margery Jourdayn.

London Bridge. Jourdain suffered the usual fate for witches at the time, burning.

Women in the sixteenth century, as today, made up only a small percentage of those convicted of crime. Then, as now, women were unlikely to be indicted for violent crimes and more daring thefts. If any were convicted of murder it was most likely that the victim was a friend or family member, and the setting was likely to be domestic rather than a street brawl. However, some crimes were particularly associated with women – for example, infanticide and witchcraft.* In the period between 1550 and 1750, 219 people were convicted of witchcraft in Essex but only 23 were men.† King James had a notable obsession with witchcraft and witches were prosecuted at a particularly alarming rate during his reign.

Burning was also used as punishment against religious heretics. For example, Joan of Arc, depicted in *Henry VI* Part I, was famously burned to death at the Vieux-Marché in Rouen. The play does not depict the burning itself (far too risky in a wooden building), but it is reported by other characters. As an act of mercy they say Joan's death is to be hastened by adding barrels of burning pitch to the pyre.

The experience of death by burning could differ enormously depending on many variables. Sometimes

---

* Lady Macbeth is an example of a Shakespearean character who may have been guilty of infanticide. The text is vague and has been a rich source of speculation over Lady Macbeth's character and motives for her actions.

† Witchcraft was also extremely regional as well as gendered. Essex seems to have had a particular problem with witches and has far more convictions than other parts of England.

the condemned died relatively quickly; on other occasions their agonised screams could be heard for a long time as their bodies were slowly consumed by the flames.

Damage to the body by heat depends not only on the temperature but also on how long the tissues are exposed to this heat. Above 42°C (108°F) human cells start to self-destruct. As cells die, particularly brain cells, control of critical functions is lost. The minimum temperature that can cause damage to skin is 44°C (111°F) but it requires at least five hours of exposure at this temperature for a burn to appear. At 60°C (140°F) only three seconds of exposure will produce a burn.

Human bodies are well adapted to maintain a fairly constant core temperature. Veins near the surface of the skin dilate to increase blood flow and radiate heat. Sweating cools the skin as moisture evaporates and layers of fat act as insulation to buffer sudden changes in external temperature. Using these basic systems, the body can survive for a few minutes at temperatures over 90°C (194°F). In extreme conditions, the changes occur too quickly or the heat is too intense for the body to manage.

In fires, the temperature is obviously high, but it might not be the heat specifically that causes death. The inhalation of toxic gases – carbon monoxide, and to a lesser extent cyanide – can poison the body by disrupting the oxygen cycle. In slow-smouldering fires, carbon monoxide content in the smoke tends to be very high.

In intense fires the production of carbon dioxide (a non-toxic gas) can displace the oxygen in the air, causing asphyxiation. Also in very intense fires there may be thermal damage (burning) from hot gases entering the air passages and lungs. The high temperatures of the flames cause the skin to shrink and split, exposing the underlying fat, which may burn. Pressure inside the skull can build as the contents are heated, resulting in skull fractures.

Women's bodies tend to burn faster than men's because of the difference in fat content. But all human bodies take a long time to burn and high temperatures to reduce them to ashes. In real life, Joan of Arc's burnt body was shown to the crowd to prove she had not escaped the flames. The remains were then burned again, twice, before her ashes were scattered in the Seine.

★ ★ ★

All forms of execution in the sixteenth and seventeenth centuries were exceptionally brutal and likely to cause considerable suffering. Even beheading, the punishment for traitorous nobility, seen as a considerable mercy compared to hanging, drawing and quartering, was not the swift, pain-free experience many thought it was.

Beheading was carried out by axe blows to the back of the head. The neck contains strong muscles, the spinal cord, protected by bone, and the trachea, ringed by strong cartilage; all will give considerable resistance to the blade of an axe. It required a skilled, strong man and a sharp blade to make it through the neck in one blow. Horrific tales of multiple blows being needed were all too common.

In the politically turbulent times depicted in Shakespeare's histories, there were many beheadings of nobles who dared to cross their king. A large number of characters have their last stage appearance as a disembodied head. Seven plays require the use of prop heads and three of them are used in *Henry VI* Part II, two of which, belonging to Lord Saye and Sir James Cromer, are separated from their bodies by the order of Jack Cade, whom we met earlier.* The two heads are stuck on poles and made to kiss for the entertainment of the crowd. Shakespeare and his fellow playwrights were not unusually bloodthirsty for the time they were writing in. Numerous fake heads would have been stacked in the props cupboard of every Elizabethan and Jacobean acting company.

It may be no surprise that it is lack of oxygen that causes death in decapitation, but it comes via two mechanisms – severing of the nerve signals to and from the brain, and rapid blood loss. Cutting the spinal cord means that signals from the brain are disconnected from the muscles in the chest that enable breathing. Oxygen can no longer be replenished and lack of oxygenated blood will quickly stop the heart.

The neck contains some of the largest blood vessels in the body and severing them results in major, rapid blood loss. If a third of the body's blood volume is lost, death is attributed to acute blood loss or exsanguination. Rapid loss of oxygenated blood to the brain will quickly cause unconsciousness followed by brain death. Oxygen circulating in the brain from the victim's last breath will quickly be used up. Higher functions, thought and

* Shakespeare has muddled the names. In real life it was Sir James Saye and William Cromer who were executed by Cade's rebellion.

conscious actions, will die first. The 'lower' parts of the brain, such as the medulla and brain stem, last a little longer as the body tries to survive as long as possible against oncoming death. These areas control basic functions such as breathing, but with no body connected to the brain any signals sent from the brain to the lungs will not reach their destination.

Experiments on rats show that reserves of oxygen mean they can maintain consciousness for around four seconds after decapitation. At best a separated human head might have 15 seconds before passing into unconsciousness. Between four and eight minutes later there would be complete, irreversible brain death.* There are many reports of lips and eyelids continuing to move after heads have been separated from their bodies, and much of this will be reflexes or nerves firing, making their last gasps before death. The spectacle of faces grimacing, or apparently praying in the case of Mary, Queen of Scots, would have been unnerving for spectators. From 1606 comes the legendary, but almost certainly invented, story of Sir Everard Digby, executed for his part in the Gunpowder Plot.† Digby was first beheaded and then his heart was cut out and held up before the crowd by the executioner, who exclaimed, 'Here is the heart of a traitor.' 'Thou liest,' came the clear response from the head.

---

* Medical opinion varies on this.

† In 1605 a group of conspirators planned to blow up the Houses of Parliament when King James was visiting. The plot was foiled when one of the conspirators, Guy Fawkes, was discovered making final adjustments to the 36 barrels of gunpowder that had been placed underneath the building.

The heads of the most notorious traitors were prominently displayed on pikes in several major cities. The system was simultaneously barbaric and bureaucratic. The heads of Lord Scrope and Sir Thomas Grey, executed for their plot against Henry V, had to travel through eight counties, from Southampton to York and Newcastle, before being stuck on spears above the cities' gates. The King himself therefore had to write eight letters ordering the sheriffs to permit the passage of the heads through their county. Shakespeare dramatised Scrope and Grey's plot and sentencing in *Henry V* but skipped over the tedious paperwork.

It was much easier to have the heads displayed closer to where they had been separated from their body. This meant that London, as a prominent site of execution, had an abundance of heads for public display. These were usually placed above the gate on the south side of London Bridge, the Great Stone Gate. In fact, the number of heads, and the high turnover, meant that a Keeper of the Heads was employed to prevent London Bridge becoming overwhelmed with traitorous body parts. Older heads were removed and sometimes returned to family members, but usually they were tossed over the side of the bridge into the Thames.

How long the head remained intact and recognisable depended on whether or not efforts were made to preserve it. Bacteria and birds would have made short work of anything placed on a spike fresh from the chopping block. Sometimes the head, and other body parts, might be preserved by salting, par-boiling or dipping in tar. Sometimes weather conditions helped and cold, dry winters would slow down bacterial activity that would normally cause rapid decomposition, while hot, dry summers would help to mummify the head to a certain extent. The head of Saint John Fisher, Bishop of Rochester, executed in 1537, was displayed for two weeks above London Bridge in the middle of summer without any apparent signs of decay. The well-preserved head soon started drawing crowds and speculation that its incorrupt state was a sign of his innocence. The head was discreetly thrown into the river.

Two of the heads that were displayed on the Great Stone Gate in 1583 may have belonged to Shakespeare's relatives, John Somerville and Edward Arden, who had been executed for a plot to kill the Queen. Another head, placed on the bridge in 1582, may have belonged to an acquaintance of the playwright from his childhood. Thomas Cottam, a Catholic priest, was one of the 200 Catholics to be executed during Elizabeth's reign. Thomas was the brother of John Cottam, schoolmaster at the King's New School in Stratford, who may have taught Shakespeare.* However, by the time Shakespeare arrived in London the heads may well have been dumped in the river and at the very least would have been reduced to skulls.

⋆ ⋆ ⋆

Hanging, drawing and quartering, decapitation and burning were the rare exceptions to the most common form of execution, simply hanging. Hanging was reserved for those not born to nobility who were convicted of any

* John Cottam resigned from his teaching post a month after his brother was arrested.

felony that wasn't treason or poisoning. Elizabeth I's reign (1558–1603) saw 6,160 people hanged at Tyburn alone.

In Shakespeare's plays, the majority of executions are beheadings, because the playwright mostly depicted the lives of royals and nobles. But there is one particularly memorable hanging: that of Bardolph in *Henry V*. He is a recurring character who first appears in the *Henry IV* Part I as one of Falstaff's group of thieves and pranksters that the young Prince Henry likes to spend time with. In *Henry V* he has been recruited into the King's army as a foot soldier. When the troops are marching between Harfleur and Agincourt he succumbs to temptation and steals a 'pax' from a church.* The theft is discovered and the pax returned, but Bardolph must be punished:

> For he hath stolen a pax, and hanged must a' be:
> A damned death!
> Let gallows gape for dog; let man go free
> And let not hemp his wind-pipe suffocate:
> But Exeter hath given the doom of death
> For pax of little price.

Hanging would be the normal punishment for stealing such a small item but commanders often turned a blind eye to looting by troops marching through enemy territories. Someone who stole from a church, however, might be made an example of. The affair in *Henry V* shows the strict discipline the King expected of his army. Shakespeare made use of a real event, but attributes it to a fictional character, Bardolph, a friend of Henry's from

* A pax, pyx or pix, a small lidded box used for carrying the consecrated host.

his wayward youth, to highlight his drastic change in behaviour once he assumes the throne. There is no indication in the text that the King gives a second thought to his former friend.

In real life the hanging would have been improvised using whatever rope and tree were available. But it wouldn't have been very different to public executions carried out with full legal process in England's capital. In the sixteenth century, and for a few centuries afterwards, there was no standard rope, knot or length of drop that was required for judicial hangings. In fact, there was no real science applied to the executions at all, sometimes resulting in farcical scenes.*

The approach to the gallows and the knowledge of impending death is likely to have brought about a very physical reaction in the condemned. As adrenaline rushed through their body there would be shaking, maybe even convulsions, paralysis and fainting, and they would have sweated profusely, making the job of the hangman more difficult. In the hot summer months, when sweating was exacerbated, if the knot was not tied properly, the condemned man could literally slip through the noose.

Before 1892 all hangings were carried out by the 'short drop' method. The condemned stood on a cart or a ladder to support them while the rope was tied around their neck. The cart or ladder was then removed and they were left suspended. The criminal dangled at the

---

* In the early 1550s a man was to be hanged but the rope broke under his weight. Another rope was fetched and a second attempt made, but the rope broke again. Rather than undergoing a third attempt, the man escaped with his life.

end of the rope until they died. How long this took varied. The condemned person's own body weight and gravity slowly killed them, possibly with the help of friends and relatives hanging on their feet. Temperature, length of the rope, type and position of the knot could also make a big difference. For example, usually the knot was tied behind the ear, but it often slipped. Alternatively it could be tied in front of the voice box to prevent the condemned from crying out, but this prolonged death.

In extreme situations, such as the pain of being suspended by the neck, the brain can flood the body with chemicals that effectively shut it down in an effort to preserve life. Hanging bodies might stop twitching and jerking but they were not necessarily dead.

The cause of death in short-drop hangings might seem obvious: lack of oxygen reaching the lungs. In fact there are several factors that can contribute to the death. Compressing the windpipe is actually very difficult as it is protected by tough rings of cartilage. But air can still be prevented from entering the lungs by other means. The tightening of the rope around the neck causes upward displacement of the base of the tongue, which can block the entrance to the trachea.

It isn't just air reaching the lungs that is a concern; changes in blood supply to the brain can also kill. Major veins in the neck, such as the jugular, are nearer to the surface than the arteries. Veins are also more easily compressed because they have lower blood pressure.*

* For comparison, the pressure needed to compress the jugular veins is 4–5 psi (pounds per square inch), 9–11 psi for the carotid arteries, and around 66 psi for the vertebral arteries.

Because of the extreme stress of the situation, the heart would be beating fast, pumping blood into the brain in an effort to preserve it. The blocked veins mean the blood cannot drain and becomes dammed up inside the head. Pressure can build up to enormous levels and the brain can effectively be pulped. Signs of what is going on in the interior of the skull can be clearly seen in the face as it darkens and swells with suffused blood. The eyes bulge and the tongue might protrude. From a contemporary medical point of view the criminal was still alive as the heart continued to beat, though, thankfully, they were probably unconscious owing to brain damage.

Although the majority of criminals expired on the scaffold, there is evidence from the nineteenth century that a considerable number did not. William Cliff's records of dissections from 1830, carried out on hanged murderers, show that there were 10 out of 35 cases where the heart was still beating after the criminal had been cut down from the scaffold. Expertise in execution methods cannot have been any more sophisticated in Elizabethan England and so it can be assumed that not everyone in Shakespeare's day died while they were hanging.

Certain signs would be looked for to ensure that dead bodies, rather than live ones, were cut down from the scaffold. These included blackening of the face and hands, cloudy corneas, lack of signs of breathing or sensibility. But not all hanged criminals displayed these classic signs. In some cases the heart clearly stopped beating long before enough pressure could build up in the head to produce these effects.

Pressure from the rope on the vagus nerve in the neck can result in the rapid onset of cardiac arrest. Stimulation of these nerve endings can trigger the 'vagal reflex', sometimes called 'vagal inhibition', 'vasovagal shock' or 'reflex cardiac arrest'. Rapid cessation of the heart's action means there is no time for the accumulation of blood in the head to produce the blackening and bulging of the facial features. However, this reflex effect is much more common in cases of manual strangulation than in hanging (see Chapter 5). Of course, death can be a result of several factors acting together to varying degrees.

In the final stages before death the automatic nervous system is activated and foul-smelling excrement is expelled. After death, a person left hanging for some time will have a much paler torso but the hands, legs and feet can be very dark, almost purple, because of gravity pulling the blood down and pooling in the extremities. Bardolph's hanging body would have been a grisly, stinking warning to Henry V's troops as they marched past it.

★ ★ ★

If the sight of a hanging wasn't macabre enough to scare the populace into good behaviour, still worse punishments were available. Hanging in chains, or gibbeting, was used for more serious felonies such as wilful murder committed with premeditated malice or during a notable robbery. The felon was hanged alive in chains near the scene of the crime, left there to starve and for his body to rot: 'Upon the next tree shalt thou hang alive, / Till famine cling thee' (*Macbeth*). If the judge felt lenient they might allow the condemned to be first strangled with a rope before his body was left on display. In the winter months cold weather may have hastened proceedings through exposure to the elements. The chains were constructed like cages to hold the remains in a standing position. The body might have been further bound together to prevent it from literally falling apart, or dipped in pitch or tar to preserve it for longer.

In *Antony and Cleopatra*, Shakespeare uses gibbeting to show the extremes the Egyptian Queen is prepared to go to rather than subject herself to Roman rule: 'Rather make / My country's high pyramids my gibbet / And hang me up in chains!'

In many cases of gibbeting the convicted starved to death or, more likely, died of dehydration after a few days. Dehydration means there is not enough water in the body to carry out normal metabolism. When a person is deprived of water, death can be expected within 10 days, or less if the ambient temperature is high. If there is water but no food, starvation will cause death. The time it takes to starve to death varies depending on the relative fitness and fatness of the victim but is likely to be between 50 and 60 days, as long as there is adequate water.

Starvation seems a particularly cruel way to die, but that has not stopped some people, both real and fictional, from using it as a form of execution. In 1300 Henry IV seized the crown from Richard II and had him imprisoned. As long as Richard remained alive there was a threat of rebellion against Henry and so he wished the deposed King dead.

In Shakespeare's portrayal of the events in *Richard II*, Sir Piers Exton overhears King Henry ask 'Have I no friend will rid me of this living fear?' Exton interprets this as the King hoping someone will kill Richard on his behalf, so he and two murderers go to Pomfret Castle where Richard is being held, and attack him. Richard manages to kill two of his opponents in the fight but meets his death at the end of a poleaxe brought down on his head by Exton. Shakespeare's version, based on accounts narrated in Holinshed's *Chronicles*, may have been dramatic, but it was completely false.

Examination of Richard's skull in the seventeenth century found no marks of a blow or wound. A contemporary French chronicler wrote that he was starved and left so hungry that 'Richard used his teeth to tear strips of flesh from his arms and hands and devoured them.' Others claimed he was put on a starvation diet and some say his starvation was self-inflicted. It is, however, generally believed that Richard II died by slow starvation at Pomfret Castle, whether at Henry's orders or not is less clear.

Shakespeare did use execution by starvation, but for a fictional character, in the brutal and bloodthirsty *Titus Andronicus*. In the play Aaron, a conniving and manipulative individual involved in plotting several murders and a rape, is sentenced to be buried up to the chest and starved. Compression of the chest by the surrounding sand or soil may have been enough to kill him long before starvation intervened. The chest needs a few inches to expand for air to rush into the lungs. The weight of soil on the chest can kill rapidly as accidents on construction sites have proved. When the chest is held in a fixed position, the parts of the body left exposed above the soil are grossly discoloured by blood that has been pumped up from the heart through the arteries but is unable to return as easily because of the compressed veins, and there can be copious bleeding from the ears and nose. It is an exaggerated form of slow death from strangulation. There is no mention of what happens to Aaron after he is led offstage – we don't know whether he dies a long drawn-out death from starvation or a rapid death from crushing.

★ ★ ★

Crushing, or pressing, was also a form of execution in use in the sixteenth and seventeenth centuries. Normally, the death penalty for felonies not only deprived families of loved ones, but forfeit of their possessions left wives and children destitute. There was one way of at least preserving property and possessions for those left behind. At trial, the accused could refuse to offer a plea. Without a plea the trial could not continue and no sentence could be given. The accused escaped the financial penalties of the law, but did not escape with their life. They were still penalised for their silence, by '*la peine forte et dure*', the legal expression for being pressed to death with heavy weights. The victim was stretched upon the ground and a board placed on their chest on which weights were piled until they died.

A healthy person can breathe with 180kg (400lb) on their chest for two days before getting tired, a fact discovered in 1692 in colonial America when Giles Corey was accused of witchcraft. He was sentenced to be pressed to death with 400lb of stones placed on his chest. It took him two days to die, and apparently his last words were 'More weight.'

In Elizabethan England, electing this manner of death was seen as courageous. The death was therefore hastened to minimise suffering. This was achieved by placing a sharp stone or a piece of wood under the person's back that would crush the spine and stop nerve signals reaching the lungs. The punishment was more common than might be expected. Between 1603 and 1621 at least 41 men and three women were pressed to death in Middlesex alone.

Although Shakespeare had none of his characters punished in this way, he made several references to the practice. For example, in *Richard II*, the Queen says 'Oh I am pressed to death through want of speaking.'

Shakespeare was relatively restrained when it came to executing his characters, especially onstage. But he used his knowledge of capital punishment, and played to his audience's experience of the real thing, to maximum effect. Subtle references, asides, even scaffold humour, are far more effective than gruesome simulations or detailed descriptions of executions, which the audience would have been all too familiar with in any case.

CHAPTER FIVE

Murder, Murder!

> O wondrous thing! How easily murder is discovered!
>
> *Titus Andronicus*, Act 2, Scene 2

Everyone loves a good murder. Our desire to watch, read and hear about murders in all their grisly detail seems insatiable. Crime dramas on TV are perennial favourites and news outlets eagerly report details of the latest homicide, but this is nothing new. Shakespeare knew what would attract a crowd when he created some of the best-known villains of the stage and dramatised what became often repeated, even parodied, moments of murder.

Shakespeare's tragedies and histories are littered with the bodies of characters who got in the way of someone's ambition or were cut down because of some perceived insult. Othello deliberately suffocates Desdemona, thinking she has been unfaithful to him. Macbeth clears his path to the crown by slicing through anyone who might try to take it from him. He is also careful to plant evidence on others to cover his tracks. Shakespeare portrays the Plantagenet kings bumping each other off at an alarming rate to secure their positions on the throne (as they often did in real life) though they usually got others to do the dirty work for them.

If Shakespeare and his fellow playwrights couldn't sate the bloodthirsty appetites of their audiences, there were plenty of other outlets supplying stories of murders and murderers. From the early years of Elizabeth's reign the public were treated to a constant stream of pamphlets and ballads describing the lives and eventual fates of notorious criminals. These were far from being well-researched, fact-based reportage. 'The prose and verse are largely stereotyped, while the illustrations are scarcely credible' is how one student of popular literature described them.

Murder is a fairly broad category and could reasonably include deaths that occur as a result of a violent argument or war. Some have used the chaos of the battlefield to exact revenge on a rival, as in the case of John Stafford who in 1460 at the battle of Northampton sought out Sir William Lucy, husband of the woman he was having an affair with, and killed him.* Executions could also be seen as the deliberate planning of a death, but at least there is usually a judicial process beforehand that gives the act a different legal status. To narrow things down a little bit this chapter will look at Shakespeare's accounts

* Lord Stafford makes an appearance in *Henry VI* Part III, but in a non-speaking role, where he is reported to have been slain in the Battle of Northampton by the soldiers of Henry VI.

of deaths brought about as a deliberate, planned act by another person – murder with malice aforethought – and how *not* to get away with it. Poisonings will get their own chapter.

★ ★ ★

Some of Shakespeare's murders are notable because attempts are made at forensic examination to determine the cause. Other deaths are interesting because Shakespeare was happy to lay the blame on a specific person, even if historically there was some doubt in the matter. In doing so, Shakespeare has possibly tarnished the reputations of several historical figures. The murder of Humphrey, Duke of Gloucester by the Earl of Suffolk in *Henry VI* Part II is an example of both historically inaccurate finger-pointing and surprisingly detailed forensic examination.

In the play Duke Humphrey is found dead in his bed. The body is closely examined to determine if the cause was natural or if there has been foul play. It is part forensic examination and part whodunit, all played out in a few short moments. It is almost the prototype police or detective drama, with investigation, suspicion and accusation, but this is not a Shakespearean murder mystery. The idea of a detective drama, where the reader or audience is also ignorant of the culprit and can play along to see if they guess before the characters, was invented in the nineteenth century. Shakespeare never created any mystery over who was responsible for the death; it was well advertised to the audience even if the characters in the play are ignorant. But the investigation of Duke Humphrey's body is not a million miles from modern detective dramas that show forensic experts and detectives crowding round a dead body, looking for evidence of the cause of death and who might be responsible.

Forensic science, in terms of fingerprinting, toxicology and so on, was developed in the late nineteenth century, but that does not mean that no attempt was made to determine cause of death or seek evidence of foul play before then. The role of coroner was established in England in 1194 and while the job had many responsibilities, one of the main ones was determining the cause in cases of suspicious death. Forensic dissections have been carried out in England since the thirteenth century. Shakespeare would have been well aware of this kind of examination, as his father was required to be present during post-mortems as part of his job when he was an alderman in Stratford.

In the play, the audience knows that Duke Humphrey has been murdered. When the body is discovered Warwick immediately suspects foul play: 'I do believe that violent hands were laid / Upon the life of this thrice-famed duke.' Suffolk pretends to be incredulous, knowing full well the Duke was murdered on his very specific and detailed orders. He had clearly hoped to get away with it by arranging things in such a way as to make it look like natural causes. But the murderers were either not up to the job, or Warwick's sharp eyes are too good for Suffolk's planning.*

* Suffolk gets his just deserts: he is later sent into exile but captured by pirates on his way to France and beheaded.

The first thing Warwick notices is the colour of Duke Humphrey's face. Pallor can reveal a lot about a corpse. The colour can be very different from that of a living person, but it varies enormously between corpses and even between different parts of the same corpse. For example, fluorescent yellow means liver failure. Pink strongly suggests carbon monoxide poisoning. Very, very pale is due to haemorrhage. Blue results from cyanosis. Dark and pale patches can be scrutinised to see if a corpse has been moved.

Warwick compares the colour of Duke Humphrey's face with the paleness he expects to see in a dead body: 'Oft have I seen a timely parted ghost, / Of ashy semblance, meagre, pale and bloodless, / Being all descended to the labouring heart [...] Which with the heart there cools, and ne'er returneth / To blush and beautify the cheek again.' Exactly as Warwick points out, when the heart stops pumping blood around the body, gravity causes it to pool at the lowest points. A dead body lying in a bed would be expected to have a pale face and purplish colouration (known as lividity) on the back, buttocks and back of the legs where the blood collects. Pale patches occur where pressure is exerted against the skin and the blood is squeezed out of the tissue. As the blood moves down, higher parts of the body can look flat or depressed as blood and plasma slowly sinks.

Warwick goes further: 'see, his face is black and full of blood: / His eyeballs further out than when he lived, / Staring full ghastly like a strangled man'. He is convinced the Duke has been killed by strangulation and lists the signs he considers to be evidence of this. The bulging eyes and 'blood is settled in his face' suggests something has stopped it from sinking. Pressure on the neck can squeeze veins shut, preventing blood from draining from the head.

Manual strangulation is one of three types of asphyxia,* but death is not actually caused by the compression of the windpipe preventing oxygen entering the lungs, as may be thought. The pressure needed to compress and totally block the trachea is said to be around 33 psi (pounds per square inch), or like trying to squash flat a well-inflated car tyre using just your hands. Instead, death is due to compressed veins damming up blood inside the head, or stimulation of the vagus nerves in the neck causing cardiac arrest, or a combination of these factors (see Chapter 4).

There are several clues to corroborate Warwick's theory. 'But see, his face is black and full of blood' may also refer to another phenomenon often observed in manual strangulation – petechiae. These are small red or purple spots (0.1–2mm in size) caused by the rupture of the thin walls of tiny blood vessels near the surface of the skin. Their appearance is especially common in lax or unsupported tissues, such as the eyelid, forehead or behind the ears. Sometimes the bleeding can be extreme, and not held under the skin, so blood escapes from the nose or ears. However, petechiae are not just caused by strangulation; violent sneezing can result in their almost instant appearance, and so their presence cannot be taken as confirmation of strangulation, but, along with other evidence, has to be explained.

---

* Asphyxia roughly translates as 'without pulse' but modern use of the word has come to mean 'without oxygen'.

If the Duke was strangled, as the above evidence suggests, it is strange no one seems to have checked the Duke's neck. There could be marks left by a ligature, or bruising from hands, or scratches from the victim trying to fight off his attacker, but then these are not always apparent. Natural post-mortem changes can also affect colouration and the longer the body is left lying before it is discovered, the more opportunity for post-mortem changes to occur. If the Duke had died face down, of natural or unnatural causes, blood could have pooled in the face to give the dark discolouration. In a modern court of law, without evidence of bruising on the face or neck, the colour of the face alone would not be accepted as conclusive proof of foul play. Or, perhaps the discolouration is due to bruising from an assault, and there is other evidence of a struggle that would back this up.

The lines 'his nostrils stretched with struggling: / His hands abroad displayed, as one that grasped / And hugged for life and was by strength subdued' could be a description of rapid onset of rigor mortis. During life the body produces adenosine triphosphate (ATP), a form of fuel used by the cells to carry out various tasks within the body. One role of ATP is to allow the smooth gliding of muscle fibres over one another. After death no more ATP can be produced; the fibres become bound to one another and the muscle stiffens. Muscles exhausted of their energy stores after a lot of exertion become bound up and hardened into a fixed position after death. But the exertion that caused rigor mortis in Duke Humphrey's face and hands may have been due to convulsions brought on by natural causes.

Look on the sheets his hair, you see, is sticking.
His well-proportioned beard, made rough and rugged,

Like to the summer's corn by tempest lodged:
It cannot be but he was murdered here:
The least of all these signs were probable.

The disordered beard and hair sticking to the sheets are perhaps signs that the Duke was smothered, which is the second form of asphyxia.* When the mouth and nose are obstructed, the body is both deprived of oxygen, and also cannot expel the carbon dioxide (CO_2) produced in the body by respiration. As CO_2 levels in the blood increase, the pulse quickens and blood pressure rises. The increase in CO_2 lowers the pH of the blood and creates a state called hypercarbia, which produces extreme anxiety. There will be increasingly strenuous attempts to draw breath and the victim is likely to pass out after about 15 seconds, but they are not dead.

There are reserves of oxygen within the body and some parts of the body have lower oxygen demands than others. The brain has a very high oxygen demand and very little in the way of reserves. Without oxygen, cells cannot get energy and quickly die. Cells in the central part of the brain hold out the longest, but in the space of a few minutes cell death will have reached even these protected areas that regulate breathing and heartbeat. As a result convulsions may be triggered and the efforts to breathe become increasingly weak and shallow; the heartbeat becomes irregular and finally stops.

Covering the mouth and nose may prevent oxygen from entering the lungs and circulating but dark,

---

* The third type of asphyxia is when chemicals disrupt oxygen processes within the body, such as in cyanide poisoning.

deoxygenated blood is still free to move around the head and body. This can cause a blue discolouration known as cyanosis. But if the Duke was smothered, any discolouration would be expected throughout the body, not just the face.

None of the signs Warwick lists are in themselves conclusive proof of foul play, but collectively they are certainly suggestive. The discussion around the state of the body and the circumstances of the death perhaps echoes some of the concerns expressed when the real-life Duke Humphrey was found dead in his bed in 1447.

The real-life Duke of Suffolk had accused Humphrey of plotting against the King. Humphrey denied everything but on 11 February he was placed under house arrest and 12 days later he died. According to Holinshed's *Chronicles*, his body was shown to the Lords and Commons and showed signs that he had died of palsy, or of an impostume (an abscess).

Duke Humphrey's death was certainly convenient, and some thought it was a little too convenient. Rumours started very early that the Duke had been killed by strangulation, suffocation between two beds, or even by 'a thrust into the bowel with an hot burning spit'. The speculation was that Suffolk was responsible, but there is no *evidence* that the Duke was murdered. Duke Humphrey's friend, Abbot Whethampstead, believed he had died of natural causes. He was 56 at the time of his death and his life had been filled with drink and debauchery. The Duke had lain in bed in a coma for three days before expiring and so it was probably a stroke that killed him.

The gossip surrounding Humphrey's death contributed to the belief that Gloucester was an unlucky title – three of those who were given it died miserable deaths. Thomas of Woodstoke, was the first unfortunate Gloucester. He was murdered in 1397 by being smothered with towels (although when the murder is reported in *Richard II* it is implied he was stabbed). The aforementioned Duke Humphrey was the second Gloucester to come to a horrible end. Shakespeare picked up on this story and when young Richard Plantagenet, son of the Duke of York, is given the title Gloucester in *Henry VI* Part III, he begs to have a different dukedom, as Gloucester brings bad luck. And, for the man who later became King Richard III, it certainly did.*

★ ★ ★

Something about Duke Humphrey's death in his bed by suffocation or strangulation may have sparked a particular dramatic interest in Shakespeare. He used more or less the same set-up for one of his most famous murders: that of Desdemona at the hands of her jealous husband, the title character in *Othello*.

Shakespeare's tale of betrayal and deception has the wicked Iago convince his master, Othello, that his beloved wife has been unfaithful. In a desperate rage he kills her by strangling her. He thinks she is dead but she revives long enough to tell her maid who attacked her, before collapsing again and dying.

The tale is borrowed from *A Moorish Captain*, an Italian short story by Giovanni Battista Giraldi that would almost certainly have been lost and forgotten if it

* After Richard's death no one was given the title for over 150 years.

wasn't for the Shakespeare connection. The Bard was expert at taking other people's stories and reworking them into something brilliant. In the original tale Desdemona is beaten to death by the Iago character while the Moor watches. Then the two lay the lifeless body on her bed and pull down the roof on top of her to make it look like an accident. The original story also has both men arrested and tortured, but the Moor refuses to confess and so is banished. He is eventually killed by Desdemona's relatives. The torture of the Iago character results in his body rupturing and he is 'taken home, where he died a miserable death'.

Shakespeare made some significant changes when he adapted Giraldi's tale for the stage. Changing how Desdemona is killed avoids the practical difficulties of collapsing a roof on top of someone onstage. But other changes radically alter the relationship between both Othello and his wife and Othello and Iago. Othello and Iago no longer work together and instead Othello becomes directly responsible for his wife's death. Iago's manipulation of everyone onstage has left audiences speculating over his motives for over four centuries.

In the play the stage directions say '[*He stifles her*]', which could mean Othello uses the bedclothes or his hands to suffocate her. Either method would come under the generally accepted definition of asphyxia but could potentially result in very different physical signs on Desdemona's body.

Regardless of the method he uses, Othello thinks Desdemona is dead and stops the process of suffocation: 'Ha! no more moving? / Still as the grave'. He is interrupted by the arrival of Emilia, Desdemona's maid, who fails to notice anything amiss. She is clearly distracted by news of another murder and may not notice her mistress lying on the bed. Even if she did there may have been nothing to rouse her suspicions.

In fact Desdemona is not dead and has merely fallen unconscious. Removing the bedclothes or his hands would allow the flow of oxygen into and through the body to be returned to normal. Most people can hold their breath for 30 seconds without suffering any permanent harm.* But withholding oxygen from the brain beyond a few minutes usually means brain cells start to die. Had she been successfully strangled by Othello's hands, it would be expected that her face would look more like that of Duke Humphrey, described above, with prominent eyes and discolouration of the skin. But strangulation does not always produce petechiae and there may have been no sign of bruising.

Initially, no damage seems to have been done to Desdemona's brain: as soon after Othello stops trying to kill her, she regains consciousness and begins speaking coherently. At this point she should have made a full recovery, but then she collapses again and dies, so something else must have happened to her during the assault.

The suddenness of her collapse indicates a dramatic change in health. One possible explanation is that she hit her head during the struggle with Othello. The knock wouldn't need to be particularly severe; it is where the blow occurred that is more important. Some parts of

* Divers can train themselves to hold their breath for much longer. The current record stands at 22 minutes and 22 seconds.

the skull are more vulnerable than others and the force of impact doesn't always obviously correspond to the extent of damage. Severe blows can leave an individual without concussion but a slight knock can be followed by unconsciousness and even death.

A sudden acceleration or deceleration, such as a knock against a wall or bedpost, stops the skull, but the tissue inside continues to move until it meets the walls of the skull. Effectively the brain sloshes around within the limited space available. Sudden changes in force can cause a vein to rupture and bleed into the cranial cavity, and then pressure can build and cause further damage to brain tissue. When the pressure is too high it results in sudden collapse, and the pressure on the brain from the bleed can be fatal.

Othello is a notable exception among Shakespeare's murderers in that he readily confesses to his crime. Most of the others try to justify their actions, shift the blame on to others or disguise the crime in some way. The conspirators who murder Julius Caesar do so because they see him as an unsuitable ruler. Macbeth leaves bloody daggers with Duncan's sleeping guards so everyone will think they are guilty, and he also tries to stage the murder of Banquo as a robbery. As we have seen, Suffolk tries to make his murder of Duke Humphrey look like natural causes. In *King Lear*, the murderer attempts to cover his tracks by trying to disguise the murder as suicide.

★ ★ ★

As with many of his works, there are several sources Shakespeare probably drew upon to create *King Lear*.

Holinshed's *Chronicles* provides the main character names: King Leir, said to have been King of Britain in the eighth century, and his daughters Gonorilla, Regan and Cordelia. It is also from Holinshed that he got the basic premise of a king who wants to divide his kingdom between his three daughters, but cuts off the inheritance to the youngest, and suffers the consequences. There were already several theatrical adaptations of the tale before Shakespeare took it on, so the story of King Lear was well known to Elizabethan audiences. In all the previous versions of the tale Cordelia and the King are reconciled and live happily ever after.

In Shakespeare's adaptation, Edmund gives the instruction 'To hang Cordelia in the prison and / To lay the blame upon her own despair'. Lear interrupts the guard while he is attempting to hang Cordelia. He kills him and carries his daughter's body onstage. Hoping he has reached her in time, he desperately checks for signs of life.

Cordelia's face must be pale and there can be no obvious injuries, otherwise Lear would accept her death more readily. It therefore seems likely that the murderer choked, strangled or smothered Cordelia before hanging her. As we have seen from Desdemona's revival in Othello, and in the chapter on execution, hangings and attempted suffocations were not always fatal. Lear has reason to hope that Cordelia may yet still live and everyone is kept guessing over several lines as Lear alternates between hope and despair. But there is no hope. Cordelia is dead and Lear dies from grief. Elizabethan audiences would have been expecting a last-minute recovery and the deaths would have come as a genuine shock. Edmund confesses, but if he hadn't, and even if Lear had not interrupted the hanging process, he is still unlikely to have got away with it.

There are clear post-mortem differences between strangulation, suffocation and hanging. If Cordelia was garrotted, even if it was with the same rope that was then used to hang her, there would be evidence that she had died before being suspended. There would be no need for sophisticated forensic examination; the pattern of the rope marks should give a clear indication of foul play. Ropes and other ligatures usually leave marks on the neck, and these would be at different angles depending on how the rope was used. For strangulation, the line would be horizontal and would usually circle the whole neck. In a hanging the ligature mark would be angled up towards the point of suspension with a gap underneath the knot.

If Cordelia had been suffocated, and was dead before she was hanged, the rope would not make the same marks on the skin, as her blood would no longer be flowing. However, not all cells die at the same time so there is a period of time around the point of death, referred to as perimortem, where it can be difficult to judge if injuries were obtained before or after death.

Signs of scratches at the neck might also indicate that she was trying to fight off her attacker or remove the rope, but would not be conclusive signs of murder taken on their own. But careful examination and combined evidence would make a strong case for murder.

In *King Lear* Edmund is undoubtedly a scheming, villainous character, but even he doesn't come close to Shakespeare's most notorious murderer – Richard III.

The ruthless, plotting, evil hunchback is perhaps the prototype pantomime villain.

★ ★ ★

Shakespeare was certainly happy to apportion blame where it suited his dramatic purposes and as a result several historical figures have had their reputations tarnished. Richard III has perhaps suffered more than most at the Bard's hand, though he was certainly no saint. Even his hunched back, his limp and his withered arm is a gross exaggeration.* Shakespeare, and many others, exaggerated King Richard's physical characteristics to show his inherent evilness.

Shakespeare was writing at a time when it was politic to curry favour with the Tudor Queen Elizabeth and deride her Plantagenet predecessors. In many respects he was merely repeating the propaganda of his day, with a few embellishments for dramatic purposes. That is not to say that Richard wasn't particularly ruthless, but there may not be quite so much blood on his hands as the play might lead you to believe.

Richard Plantagenet, Duke of Gloucester, appears in three of Shakespeare's plays before he is made king, which is a considerable back story. Though he appears in all three parts of *Henry VI*, it isn't until the final part that

---

* Examination of Richard's skeleton has revealed that he had a scoliosis, or curvature of the spine, that would have made his trunk appear shorter, and one shoulder slightly higher than the other, but nothing that couldn't be disguised by a good tailor. There is no evidence that he had a withered arm or would have limped.

his ambitions become clear. He covets the crown for himself, but he points out that 'many lives stand between me and home'. Not that this seems to deter him: 'were it farther off, I'll pluck it down'. He spends the rest of this play and the following one (*Richard III*) killing everyone in his way until he achieves his goal. It is certainly dramatic, just not always historically accurate.

According to Shakespeare, Richard III's deadly tally is considerable. In the fourth act of *Richard III* Queen Margaret meets with the Duchess of York and Queen Elizabeth to compare notes and tick off a list of murders they attribute to Richard. Queen Margaret sets them off:

> Tell o'er your woes again by viewing mine:
> I had an Edward, till a Richard kill'd him;
> I had a Harry, till a Richard kill'd him:
> Thou hadst an Edward, till a Richard kill'd him;
> Thou hadst a Richard, till a Richard killed him …

In total the Bard attributes 11 murders to Richard (12 if you count the mental torment he inflicted on his brother Edward IV as a contributing factor in his death). He was also responsible for many deaths during the various battles he took part in, but Shakespeare gives him a few extra for good measure. For example, in *Henry VI* Part II he is shown killing Somerset during the Battle of Saint Albans, though he was in fact three years old at the time. His murderous career begins in earnest in *Henry VI* Part III, where he fights alongside his brother King Edward IV in the Battle of Tewkesbury against Henry VI's forces. Edward is victorious, but to secure his place on the throne Richard helps him eliminate any competition for the crown, and thereby moves himself closer to it as well. The most serious threat is Prince Edward of Westminster, King Henry VI's only son and heir. Edward of Westminster was just 17 in 1471 when he went into battle on his father's side. Richard was only a year older.

According to contemporary accounts, Edward of Westminster 'died in the field', the only heir to the English throne to die in battle. However, there are alternative theories as to what happened. Croyland, writing in 1486, states the Prince died 'either on the field or after the battle, by the avenging hands of certain persons'. Those certain persons were said to be Thomas Grey, 1st Marquis of Dorset, and Lord Hastings. Shakespeare adopted the more dramatic version and then went on to embellish it further. Instead of Dorset and Hastings, he has the Prince stabbed by Edward IV and his two brothers Richard and George. Edward of Westminster is killed after the fighting has ended – this is murder.

With Henry's heir dead there is now only the niggling worry that the deposed King Henry VI will become the inspiration for rebellion. Richard immediately sets off for London and the Tower where Henry is being held captive.

Henry VI may have been a vague and ineffectual king, easily swayed by those around him, but Shakespeare allows that he at least has the measure of Richard. When he appears in Henry's rooms the normally mild-mannered King, who never has a bad word to say against anyone, greets him with, 'Ay, my good lord:—my lord, I should say rather; / 'Tis sin to flatter; 'good' was little better: / 'Good Gloucester' and 'good devil' were alike'.

Henry is also under no illusion as to why Richard is there, 'But wherefore dost thou come? is't for my life?' After Richard confirms he has already killed Henry's son the King prophesies that it is the first murder of many more to come. Tired of listening to Henry run through Richard's many faults, 'I'll hear no more: die, prophet in thy speech': Richard stops him talking by stabbing him.

Richard makes little effort to conceal his crime. He may not have told his brothers exactly what he was going to London to do, but it wasn't difficult for them to guess. He may have sent Henry's guard out of the room so he couldn't witness the actual murder, but it wouldn't have been difficult to put two and two together. At the end of the scene he drags the body of the murdered King out of the room, but this is because the stage needs to be clear for the next scene and not because he is trying to hide the crime. In real life the details of how Henry VI died were at least suppressed and a cover story concocted, even though it stretched credibility to its limits.

Henry VI died, if not immediately, then not long after the battle of Tewkesbury. The official account was that he died of 'pure displeasure and melancholy' on hearing that his son had been killed in battle. But it was a particularly violent kind of melancholy brought on by the rapid application of blade against flesh, a detail left out at the time. In 1911 the body of Henry VI was exhumed and examined; the bones of the skull were found to be 'much broken'. One part of the skull still had hair attached, matted with what looked like blood.

Who was responsible for breaking Henry VI's skull is less certain. Richard was at the Tower on the fatal night, and it is difficult to explain why else he would have been there. But as to whether he went alone into the King's rooms and committed regicide, shouting, 'Down to hell, and say I sent thee thither!' as Shakespeare has it, is not clear. Whether he was personally involved or not, he certainly had motive, as it brought him much closer to getting his hands on the crown.

⋆ ⋆ ⋆

Richard III picks up from where *Henry VI* Part III leaves off. The second scene of the play has Richard pacing round the coffins of his two alleged murder victims, Prince Edward of Westminster and King Henry VI. Also in the room is Edward's widow, Lady Anne, mourning over her dead husband. In case anyone in the audience had forgotten what happened in the last play, Anne uncovers Henry's body: 'see, see! dead Henry's wounds / Open their congeal'd mouths and bleed afresh!' – confirmation that Richard was guilty of his murder.

From the twelfth century right up until the nineteenth, it was common for accused murderers to be brought to the corpse of their alleged victim and made to touch the body. It was believed that wounds really did bleed again in the presence of the murderer. The test was known as 'cruentation' or 'ordeal of the brier'. It would be an extraordinary set of coincidences if this ever really did happen. After death blood may seep from wounds for a short period of time, but after about six hours it has settled and solidified to an extent that prevents further flow. It is true that blood can re-liquefy at a later time owing to decomposition. Even then, the body would likely need a good shove to get the wounds bleeding again.

If the sight of a freshly bleeding corpse wasn't sickening enough, then watching Richard woo the widow of one of his victims over the coffin is truly nauseating. It may seem unbelievable that Richard could talk Anne into marrying her husband's murderer, but that is exactly what he does. The playwright didn't invent this for dramatic purposes either. Richard might not have killed Edward of Westminster, but he did marry his widow a year later.* In the play Richard confides to the audience, 'I'll have her; but I will not keep her long.' Anne's fate as another of Richard's victims is sealed before she even has the ring on her finger.

Later in the play, in a transparent attempt to cover up the planned murder of his wife Lady Anne, Richard deliberately spreads rumours that she 'is sick and like to die'. It is heavily implied that he poisons her. Holinshed's *Chronicles* presents a slightly different account. Richard did complain that his wife was barren and started rumours that she was sick. But this was because she probably was sick, and on 16 March 1485 she died, possibly of cancer or tuberculosis. Richard's reluctance to visit his wife on her sickbed, and his apparent indifference to her death, led many at the time to believe that she had been poisoned by him. Richard also didn't wait until his wife was dead to make it known he would quite like to marry his niece, which only made things look even more suspicious.

---

* Technically Anne and Prince Edward were betrothed and not yet married. Richard in fact did not use honeyed words to win Anne over; he effectively kidnapped the reluctant 16-year-old and kept her prisoner.

Richard's treatment of Lady Anne is certainly in keeping with his villainous character, but it does not advance him any closer to his goal.* His brother George, Duke of Clarence, stands between him and the throne; being the older brother he would inherit before Richard. Shakespeare rewrote the real-life downfall of Clarence to add another murder to his growing list and to cast Richard as a supreme opportunist and manipulator.

The plan is set up in the very first scene of *Richard III*: 'Plots have I laid, inductions dangerous, / By drunken prophecies, libels and dreams, / To set my brother Clarence and the king / In deadly hate the one against the other'. Richard then bumps into Clarence on his way to the Tower after he has been arrested. Not suspecting that Richard is behind the plot, Clarence speculates that the reason Edward wants him imprisoned is because 'a wizard told him that by G / His issue disinherited should be; / And, for my name of George begins with G, / It follows in his thought that I am he.'

According to Holinshed's *Chronicles*, Edward IV did receive a prophecy that the person who reigned after him would have a name beginning with the letter G. Edward interpreted this as his brother, George, Duke of Clarence. The prophecy was later proved to be right, as it was Richard, Duke of Gloucester, who succeeded to the throne. Edward simply picked the wrong G. Shakespeare took an existing story of a prophecy and attributed it to a plot by Richard, who he then has 'urge his hatred

---

* Shakespeare does not stick to the true chronology of events. Anne lived several years after Richard became King and she became Queen.

more to Clarence, / With lies well steel'd with weighty arguments'.

In the play, Edward IV issues an order for the execution of Clarence as a traitor, but then has second thoughts and another order is given out to cancel the execution. Richard knows about both orders but acts on the first one. He instructs two murderers to kill Clarence, who plan to 'Take him over the costard [the head] with the hilts of thy sword, and then we will chop him in the malmsey-butt in the next room.'*

Stabbing followed by drowning in a barrel of wine wasn't just an example of some very black humour on Shakespeare's part; it may have some truth to it. Clarence was executed for treason on 18 February 1478, though Richard was not behind the plot as Shakespeare would have his audience believe. And some contemporary reports state that Clarence really was drowned in a butt of malmsey, some say at his own request. Drowning in alcohol is less fun than it may seem. Death occurs because the body is deprived of oxygen, and would occur long before the effects of alcohol could numb the horrible experience.

In the play everyone believes Clarence has been pardoned, so when the death is announced it comes as a terrible shock, particularly to the King, who believed the execution order had been revoked. The news of the death badly affects the King's already fragile health and he starts to doubt the Queen's motives towards him. He blames himself and also blames his wife for urging him on. Richard has removed someone who stands between

* Malmsey is a fortified sweet wine, for example Madeira.

him and the crown, and now all he has to do is sit back and wait in the hope that the King will die while his son is too young to rule and he can take over. He has done everything he can to stack the odds in his favour. Richard only has to wait until the next scene for his plan to work.

In reality Edward IV died on 9 April 1483, five years after Clarence's execution, probably of over-indulgence, and possibly from venereal disease. His 12-year-old son, Prince Edward, became King Edward V, but he never reigned. Richard had a short window of opportunity to wrench control from the Queen and her powerful family. He made the most of it. Within three months it would be Richard III, and not Edward V, who was sitting on the throne. His first move was to get custody of the young King and his younger brother Richard of Shrewsbury.

As depicted in the play, the two boys were travelling to London for the coronation when they were intercepted by Richard's men. At Stony Stratford the princes' escorts, Earl Rivers, Sir Richard Grey and Sir Thomas Vaughan, were arrested and sent to Pomfret Castle. When the widowed Queen Elizabeth hears the news she sets the scene for what is to come:

> Ay me, I see the downfall of our house!
> The tiger now hath seized the gentle hind;
> Insulting tyranny begins to jet
> Upon the innocent and aweless throne:
> Welcome, destruction, death, and massacre!
> I see, as in a map, the end of all.

Richard placed the two boys where he could control them: 'Your highness shall repose you at the Tower: /

Then where you please, and shall be thought most fit / For your best health and recreation.' Although the audience knows full well he is up to no good his behaviour towards the children is not unreasonable at this stage. Being sent to the Tower did not mean imprisonment or certain death, far from it. The Tower, as well as rooms for condemned criminals, also had royal apartments, and a well-defended castle was the safest place for the two young boys. But it was also a very convenient place to keep them if your intention was to take the crown from them.

Richard now accelerates his murderous campaign. At a meeting held at the Tower, Richard suddenly turns on his erstwhile friend and supporter, Lord Hastings: 'Thou art a traitor: / Off with his head! Now, by Saint Paul I swear, / I will not dine until I see the same.' Hastings is immediately taken away to be executed and in the following scene, just in time for dinner, his decapitated head is brought onstage to be shown to Richard.

Shakespeare's depiction of the meeting and Hastings' rapid change in fortunes is remarkably accurate. On 13 June the two did attend a meeting at the Tower, which started off well enough but ended with Hastings being beheaded on the green outside. According to the history books the execution was carried out so quickly, 'without any process of law or lawful examination', that there was no executioner's block available and Hastings had to rest his head on a log.

Rivers, Grey and Vaughan, the princes' escorts arrested at Stony Stratford back in April, were executed at Pomfret on 25 June, without trial. According to Holinshed the prisoners weren't even allowed to make the customary speeches on the scaffold. The following day Richard rode in state to Westminster Hall where, sitting on the marble throne, he took the royal oath. He was now King Richard III. There was only the small problem that the legitimate king and heir to the throne were still alive.

Shakespeare makes Richard's intentions clear early on. As soon as he has persuaded Prince Edward to go to the Tower he remarks to the audience, 'So wise so young, they say, do never live long.' Eight scenes (and in reality two or three months) later the two Princes are dead. Richard, now King, doesn't want to get his hands dirty and so he asks his page, 'Know'st thou not any whom corrupting gold / Would tempt unto a close exploit of death?' On the page's recommendation, Sir James Tyrell is employed to kill the two boys, but Tyrell subcontracts the work out to 'Dighton and Forrest, whom I did suborn / To do this ruthless piece of butchery'. Audiences are spared the sight of the actual murders but Tyrell reports the two boys are smothered in their beds.

The play follows the account of the murders described by Sir Thomas More. His emotional and elaborately detailed version of events was written several decades after they occurred. Contemporary accounts are rather vague about exactly what went on in the Tower. Dominic Mancini, an Italian friar visiting London in 1483, wrote that the two children 'were withdrawn into the inner Apartments of the Tower proper and day by day began to be seen more and more rarely behind the bars and windows, till at length they ceased to appear altogether.' Mancini suspected that the princes had been murdered but admitted that he did not know for certain. Other contemporary accounts tend to agree with Mancini's suspicions. Once the princes disappeared, it was fully expected that they would be killed and Richard was the prime suspect.

The fifteenth century may have been a violent time with political assassinations and wars bringing a high death toll, but the deliberate murder of innocent children was as shocking then as it would be today. Richard's popularity, already low, would have suffered considerably if he was proved to have been responsible for the deaths of a young king and his brother, who were also his own nephews. If he was responsible he was also sensible enough to cover up the crime.

In More's version of events, after a very detailed account of the murder, he goes on to say that the bodies were buried at the foot of a staircase, but they were subsequently re-interred in a site more fitting, on Richard's orders. Unfortunately no one knows where this more fitting site might be. Shakespeare is even more vague than More about what happened to the bodies. He has Tyrell explain to Richard that 'The chaplain of the Tower hath buried them; / But how or in what place I do not know.'

In 1674, during the demolition of a staircase leading to the chapel of the White Tower at the Tower of London, the skeletons of two children were discovered in a chest buried 10 feet deep. Contemporaries were quick to reach the conclusion that they had found the remains of Edward V and his brother and they were later interred in a magnificent urn in the Henry VII Chapel in Westminster Abbey.

Although the discovery matches More's description of the bodies being buried at the foot of a staircase, 10 feet deep is an extraordinary depth to bury evidence of a crime. It also completely disregards More's assertion that the bodies were later moved to another more fitting site. It is a remarkable coincidence but hardly conclusive proof that these were the bodies of the two princes and that Richard had killed them. With over 1,000 years of bloody history, the Tower has undoubtedly accumulated a few skeletons in its closets. The 1674 demolition may have unearthed a crime scene, but not necessarily the crime it was attributed to.

In 1933, in an attempt to answer some of the many questions over the whole affair, the skeletal remains were removed from their great urn for examination. Lawrence Tanner, archivist of the Abbey, was present along with William Wright, one of the leading anatomists of his day, and George Northcroft, president of the Dental Association. The conclusion they reached after examination of the bones was that they really were the remains of Edward V and his brother. Furthermore, there was reasonable possibility that the story of suffocation told by More was true. However, since then considerable doubt has been cast on the reliability of their report.

From the start it was assumed the two skeletons were those of the princes and the examination sought to confirm their beliefs. A more scientific approach would be to examine the bones as they were, then use those measurements to determine age and sex, and from that decide if they matched the two princes. No attempt was made to identify the sex of the skeletons, though this can be difficult to establish in pre-pubescent humans.

Most of the investigation focused on an examination of the skulls, looking for evidence of suffocation. Such evidence may not always be obvious even when examining the recently deceased, let alone 500-year-old skeletal remains. Unless the force applied to the face in the moment of suffocation was enough to break fragile bones around the nose or eyes, it seems there would be little evidence to find. Any damage that is observed also has to be carefully checked in case it occurred when the bones were removed from their grave or transferred into the urn.

The report has been re-examined and re-appraised many times since 1933. Modern conclusions vary. The lengths of certain bones and teeth are consistent with the age of the princes (twelve and nine), and the differential between the two sets of bones would appear to be correct for the two brothers. However, as you might expect, there are large margins of error. Opinion is divided as to whether the data recorded by Wright's team can be used to determine if the two skeletons are in fact related.

If modern techniques were to be applied to the remains, such as radiocarbon dating, it might give a better indication of whether they are remains from the fifteenth century or from a much earlier date. Even if DNA analysis could prove that the remains really are those of the two princes, unless there is obvious damage to the skeletons that would have caused their death, there will be little evidence of how they died, let alone who might be responsible.

Shakespeare was in no doubt that it was Richard who was responsible for the deaths of the two princes. However, the only thing that is certain is that the two boys disappeared after they entered the Tower. Any evidence of their deaths, and who was responsible, is circumstantial. Having said that, the deaths of the two boys in the summer of 1483 on Richard's orders does seem to be the most likely scenario.

Richard III was never arrested or tried for his crimes, but he didn't escape justice completely. Richard's short reign was unpopular and punctuated by rebellions. Things came to a head in 1485 when an army was raised by Henry Tudor. He had no stronger claim to the throne than Richard but he still managed to persuade several lords to come over to his side. Richard as the reigning monarch should have been easily able to raise a large army to crush Henry, but many of those he might have expected to rally to his side, didn't. Most weren't quite brave enough to join the opposition either and simply stayed out of things to watch what happened.

Henry and Richard's forces met at Bosworth. In the play, the night before the battle, all 11 of Richard's alleged murder victims make another stage appearance as ghosts.* They come to his tent, recount the crimes he has committed against them and condemn him to death, chanting 'despair and die'. Shakespeare would have felt it necessary for his audience to see Richard face up to his crimes, even if his feelings of guilt are only momentary. The next morning he shakes off his sense of foreboding and rallies his troops for the coming battle.

---

* Richard was definitely responsible for five deaths that could be considered murder and was probably behind four others. The remaining two deaths, to get up to Shakespeare's 11, were nothing to do with Richard and the Bard simply invented his guilt for dramatic purposes.

The battle itself is summarised by two brief scenes: one showing Richard's forces are losing, where Richard desperately cries the well-known lines, 'A horse! a horse! my kingdom for a horse!', and the following scene where Richard and Henry fight one-on-one and Richard is killed.

In real life the fighting lasted around two hours and by eight o'clock in the morning approximately 1,000 people were dead. Richard had fought bravely but was surrounded by his enemy's troops and hacked to death. He was the last King of England to personally fight in a battle. His body was recovered from the field and buried with little ceremony.* The Plantagenet dynasty ended and a new era of peace and prosperity began with the crowning of Henry Tudor as King Henry VII.

There was a powerful sense of justice in Shakespeare's time. It was truly believed that murderers would always be found out. Even the most successfully hidden crimes, or those whose perpetrators were beyond the reach of the authorities, could not escape divine providence. The truth would always be revealed. Shakespeare portrays Richard's defeat and death at the Battle of Bosworth as his just deserts. It would never have occurred to Shakespeare to write a whodunit, but in all of the Bard's bloody murders, as in all good detective novels, the murderer never gets away with it.

---

* Richard's notoriety grew, but his physical remains were soon forgotten until they were rediscovered under a car park in Leicester in 2013. Examination of the skeletal remains confirmed the brutal nature of his death. He was reburied in Leicester Cathedral in 2015.

CHAPTER SIX

The Dogs of War

All the battles wherein we have fought ...

Coriolanus, Act 1, Scene 6

From kings to common soldiers and emperors to ordinary folk, the majority of Shakespearean deaths come at the end of a sharp weapon.* Many of his plays are punctuated by wars and battles. Even his comedies contain the odd skirmish and swordfight.

Elizabethan audiences would have been delighted to relive past military victories and watch expert displays of swordsmanship. When plays weren't being acted, theatres often hosted fencing displays, and many in the audience would have been extremely knowledgeable and able to watch over the proceedings with a near-professional eye. Shakespeare gave the public what it wanted. He included more swordfights than any of his contemporary

* For a full list see the Appendix.

playwrights, 22 in *Henry VI* Part I alone. He mocked the tradition of duelling in *Twelfth Night* and *The Merry Wives of Windsor*;* showed street fights, like those in *Romeo and Juliet*; and represented full-scale wars in *Henry IV, V, VI* … in fact, most of the history plays and quite a few of the tragedies too.

Actors in Shakespeare's era were expected to be multi-talented. Aside from acting, they were often skilled singers, dancers, acrobats and swordsmen. Several actors were noted for their particular skill at swordsmanship (Richard Tarlton, the most famous clown of the era, was also a Master of Fence). The first fencing school in England had opened in London in 1576, only a few years before Shakespeare's arrival in the city. Many actors, and possibly Shakespeare too, trained at the fencing school run by an Italian named Rocco Bonetti in Blackfriars. The school, and the art of fencing, rapidly acquired upper-class patrons. Another Italian fencing master of the time was Vincentio Saviolo, a protégé of the Earl of Essex, who published one of the first fencing manuals in English.

Italians dominated the fencing schools in London and the Italian style was beginning to supersede the traditional English style of sword-fighting. Italian swordsmanship had a preoccupation with timing and certain defensive positions. The lunge was an important move in the Italian school but some English experts preferred a gathering step to bring the opponent within striking distance. In *Romeo and Juliet*, Romeo's side fights in the

* Duelling was briefly fashionable among the English gentry, though it never enjoyed the popularity it did in France.

English style, while Tybalt's is using the new fancy Italian techniques, much to the disgust of Mercutio. Elizabethan audiences would have appreciated Mercutio's snide remarks about Tybalt's style of swordplay.

> He fights as you sing prick-song, keeps time, distance, and proportion; rests me his minim rest, one, two, and the third in your bosom: the very butcher of a silk button, a duellist, a duellist; a gentleman of the very first house, of the first and second cause: ah, the immortal passado! the punto reverso! the hai!

The 'passado' and 'punto reverso' were genuine moves in the Italian style of rapier play, but the 'hai' is a Shakespearean invention.

Duels and street fights were easy to perform in a theatre. Even 'ordeal by battle', a judicial combat to determine both criminal cases and civil disputes, could be staged, as Shakespeare almost does in *Richard II* (Richard calls off the fight at the last minute, no doubt to the disappointment of the audience, prejudicing them against the King from the start). Recreating a full-scale battle onstage, on the other hand, was not so easy.

Several important battles feature in Shakespeare's histories; they were the turning point in the lives of many English kings. Staging them was difficult but to leave them out would be ridiculous. Fireworks and other noisy special effects could depict the chaos of war; for example, '[*Alarum as in battle. Enter, from opposite sides,* CORIOLANUS and AUFIDIUS]'.* There was also the

---

* The stage direction 'alarum', an archaic term for 'alarm', occurs over 70 times in Shakespeare's works.

opportunity for rousing speeches. Up until the middle of the twentieth century schoolboys in England often learnt Henry V's Saint Crispin's Day speech by heart. Elizabethan and Jacobean audiences would loudly cheer the English and boo the French onstage, producing an incredible effect. Tense negotiations between leaders outside the walls of besieged towns and castles could be dramatically staged using the galleried area above the stage in place of the battlements.

But showing pitched battles onstage was simply not possible. Instead, playwrights often placed the main action offstage and events are reported back. Small skirmishes and one-on-one fighting between the main protagonists would suffice to entertain the crowd and summarise how the battle had progressed: 'we shall much disgrace / With four or five most vile and ragged foils, / Right ill-disposed in brawl ridiculous, / The name of Agincourt' (*Henry V*). The bloody aftermath of battle might be shown – an excellent excuse to strew the stage with blood and body bits. And to give an impression of the scale of battle, catalogues of the names of the dead could be read out.

★ ★ ★

Battles seem to be the markers by which we see the progress of history; they tend to be the focus point. But in reality, in the medieval period, battles were few and far between. For example, the period of the Wars of the Roses (between 1455 and 1487) is covered by four Shakespeare plays: the three parts of *Henry VI* and *Richard III*. Over this 32-year period the total amount of time spent on campaign was about one year, and the total amount of time spent actually fighting was 13 weeks at most.

Wars could be waged using other tactics than battles, and land could be gained by other means. Crops could be burned, towns laid waste and castles besieged. These are difficult to stage in a theatre, but the devastation could be described to the audience as it is in *Edward III*:

> For so far off as I directed mine eyes,
> I might perceive five cities all on fire,
> Corn-fields and vineyards burning like an oven;
> And, as the reeking vapour in the wind
> Turn'd but aside, I likewise might discern
> The poor inhabitants, escap'd the flame,
> Fall numberless on the soldiers' pikes.

Full-scale battles had their place but they were often seen as the decisive end point or culmination of a war. They were big, dramatic, loud and deadly, but they were usually over in a matter of a few hours and rarely lasted longer than a day. They were likely to be decisive and devastating. There was a lot to lose, and not just the lives of the men who fought; the right to govern vast regions and populations could be at stake.

With so much to lose it seems incredible that any battles were actually fought. Leaders in this period of history often seemed to go out of their way to avoid major set-piece battles, and there was no shame in this. In *King John* there is a dispute over the English crown. The English say it belongs to John, the French say it belongs to Arthur. Both sides wish to avoid fighting and so they ask the people of Angiers to decide, but they refuse to name their king. Battle seems inevitable; however, a citizen of Angiers proposes a marriage alliance between France and England, and war is avoided. The Bastard, who had been hoping to profit from the battle, comments sarcastically that his king has been talked out of 'a resolved and honourable war, / To a most base and vile-concluded peace.'

Sometimes, after a breakdown in negotiations, or when two sides had reached an impasse, battle became inevitable. In *King John*, a further falling-out between France and England cannot be overcome through negotiations and the two sides go to war. Pride, obstinacy and underhand tricks could also play a role in battles. For example, in *Henry IV* Part I Shakespeare depicts Henry IV suing for peace with Harry Hotspur, but the message is not delivered (see Chapter 4).

Despite the obvious drawbacks, there were also advantages. In a time when there was a strict chivalric code that dictated behaviour in all aspects of life, but particularly in time of war, one death-or-glory battle would garner extreme prestige for the victor, and a noble and honourable death for the loser. Recruiting, maintaining and feeding a large army, moreover, was ruinously expensive.* One big battle could cut these ongoing expenses.

★ ★ ★

* Unlike in the play, Henry V's campaign was not easily funded by the Church. Henry not only had to pawn the crown jewels, but he also sold any spare possessions from the royal households that would fetch a reasonable price.

It is unlikely Shakespeare had any personal experience of large-scale battles to draw on when writing his plays, but there were plenty of former soldiers who had returned from wars abroad that he could have met and talked to on the streets of Stratford and London. His historical information came from written accounts such as the aforementioned Holinshed's *Chronicles*, but also Edward Halle's *The Union of the Two Noble and Illustre Families of Lancaster and York* and Samuel Daniel's *The Civil Wars between the Two Houses of Lancaster and York*. Chroniclers wrote many volumes on war, often glorifying it in the process, by detailing individual acts of valour and describing inspiring events on the battlefield. But what they recorded was far from an impartial, accurate account of what actually occurred.

The battle itself would have been chaotic and potentially spread out over a large area. It would have been difficult to keep tabs on everything. Although nobles could be identified by their armour and banners, ordinary soldiers were not always provided with uniforms and would have been difficult to distinguish from the enemy. Added to that, many chroniclers of the time were not military men and may never have witnessed a full-scale battle to understand what they wrote about – they were rarely on the spot where the action was taking place. Henry V's chronicler at Agincourt was his chaplain, and though he travelled with the army, he was not in the front line and had to make out what he could from the rear. He would also have been loyal to the King and willing to embellish and exaggerate to glorify his monarch. Stories of disease, hunger and dejection are not inspiring. Many contemporary accounts should be taken with a pinch of salt.

Shakespeare would also add his own twists to serve his dramatic interests. The playwright has Harry Hotspur killed by Prince Hal but modern historians doubt this happened. The Duke of Exeter fights at Agincourt when in fact he had been left behind at Harfleur. There are many more inaccuracies. And he wasn't always consistent in his own historical manipulations either. In *Henry VI* Part II, Clifford is killed by Richard Plantagenet (the future Richard III), but in the opening lines of *Henry VI* Part III he was 'by the swords of common soldiers slain'.

Shakespeare's plays may be lacking in historical accuracy, but he is faithful to the broad sweep of events, although he struggled to portray the scale of war onstage. The compression and distortion of huge battles into small skirmishes and individual tales can be very effective in conveying what the war was about and what it might have meant to those involved. The civil war depicted in *Henry VI* Part III is brilliantly summed up in one scene: '[*Enter a Son that has killed his father*]' and '[*Enter a father that has killed his son, bringing in the body*]'.

Individual experiences in war varied enormously. Up until the seventeenth century, kings were still expected to go into battle with their army. But their experience, alongside those of commanders and even knights, would have been nothing like that of the common foot soldier. Kings had better food, accommodation and armour to protect themselves from the worst.

Though kings and rulers could have sat out the action and directed events from a relatively safe distance, many didn't, which was a considerable risk. Foot soldiers, archers, even knights could be sacrificed and their side still be victorious. But the capture or death of a king marked the end of the battle and defeat. For example, in 1214 William the Lion, King of Scotland, went into battle with his men but his horse was brought down. Trapped under the dead or dying animal he was completely vulnerable and had no choice but to surrender immediately.

The danger of losing a king, and therefore the war, was so great that Henry IV took extra precautions. As depicted in *Henry IV* Part I, he allegedly dressed several of his noblemen in clothes and armour identical to his own. In the play, the Earl of Douglas has vowed to kill Henry but, exasperated at having killed two fake kings, promises 'I'll murder all his wardrobe, piece by piece, / Until I meet the king.' In the next scene he comes across 'Another king! They grow like Hydra's heads'. The plan works and King Henry survives the battle.

Kings could do a lot to mitigate the threats to their lives, but the ordinary soldier was not so fortunate. From the average foot soldier's point of view there may seem to be few advantages in traipsing huge distances in miserable conditions with little food, low pay and a good chance of dying – 'Would I were in an alehouse in London! I would give all my fame for a pot of ale and safety' (*Henry V*). In *Henry V*, when the King wanders through the camp the night before battle, he hears the complaints of the ordinary soldiers who sacrifice everything for their lord and run the risk of leaving their families back home destitute. One of them imagines the dead soldiers at the day of judgement: '"We died at such a place;" some swearing, some crying for a surgeon, some upon their wives left poor behind them, some upon the debts they owe, some upon their children rawly left.'

But there could be great benefits in fighting battles too. Booty could be acquired – armour, weapons and valuables would be stolen from the dead. There was also money to be made from ransoming prisoners taken during the battle. The higher the rank of the prisoner, the more money could be extorted from their estate, family and friends for their safe return. Lower-ranking soldiers fortunate enough to capture a noble or knight on the battlefield could sell them to their lord and the lord would then be able to extort the ransom payment. Everybody won, except the captive.

★ ★ ★

Of all the wars and battles that took place over the 330-odd years spanned by Shakespeare's histories, none has stayed in the collective English consciousness like the Battle of Agincourt in 1415. Shakespeare's version of events has cast Henry V as the hero figure he is still seen as today. The reality, as might be expected, was not quite the same. The Bard depicted rousing speeches, incredible acts of bravery and the undiluted success of the underdog English over the superior forces of the French. However, Shakespeare did not shy away from writing about the horrors of war.

Henry V's aim was to claim the French throne from its current occupant, Charles VI. Such an undertaking required vast numbers of troops and resources, which were slowly amassed over the year preceding the invasion. Henry landed with around 15,000 men, one of the largest armies to set foot in France since the days of Edward III nearly 70 years before. Roughly one-quarter were men-at-arms (heavily armoured and on horseback) and the remainder were mounted or foot archers.

Most of these fighting men were there to fight battles, but Henry's first engagement was the siege at Harfleur. Outside the walls of the city Henry threatens the town's governor that if he doesn't surrender he won't be able to restrain his troops:

> If not, why, in a moment look to see
> The blind and bloody soldier with foul hand
> Defile the locks of your shrill-shrieking daughters;
> Your fathers taken by the silver beards,
> And their most reverend heads dash'd to the walls,
> Your naked infants spitted upon pikes …

Threats of rape and the brutal murder of innocent citizens is not what you might expect from a heroic ruler. It gives a very different impression of Henry's character and this speech is often cut or edited for performance. But history suggests that Henry was not given to mercy and his later military campaigns were particularly brutal.

Henry terrified the town of Harfleur with more than threats. Enormous guns fired projectiles relentlessly at the city walls day after day, and miners tunnelled under them until they were reduced to rubble. Whatever structures were still standing the English then set fire to. Collapsing masonry risked the lives of civilians as well as soldiers; the dangers were well described by Shakespeare in *Henry VI* Part I.

At the siege of Orléans in 1428 the Earl of Salisbury and Sir Thomas Gargrave were wounded in the face when a cannon ball struck an observation tower they were stationed in. Salisbury was injured when a fragment of masonry 'carried away part of his face', or as Shakespeare described it, 'One of thy eyes and thy cheek's side struck off!' The face can sustain a surprising amount of injury. Wounds that look horrific will bleed profusely but unless the damage penetrates to the brain, or there is bleeding into the air passages, it is unlikely to be immediately fatal. In the play, both Salisbury and Gargrave are carried offstage still alive but are not expected to live for long. Delayed deaths can occur due to infection or complications arising from the initial wound such as thrombosis. This would appear to be the case for the unfortunate real-life Salisbury, who died 10 days after his injury. Gargrave survived only two days.

At Harfleur, those not directly involved in operating the guns and digging the mines, i.e. most of the troops, had little to do but sit around in camp among the dirt and debris that accumulated around them. By the time the French surrendered there was not much of the town left; but most of the damaging effects on Henry's troops, and the town's inhabitants, were due to dysentery rather than artillery or crumbling stonework (see Chapter 7).

It was hardly a great victory. One town, or what was left of it, had been won by the English after more than a month of effort. Henry didn't think he could make further progress into France: 'The winter coming on and sickness growing / Upon our soldiers, we will retire to Calais' and started his army marching back towards English territory. But he had to get there 'Through

France himself' and the French were looking to stop him. His army was in no fit state to engage the powerful French army, but they might not be able to avoid it. As Shakespeare put it, 'We would not seek a battle, as we are; / Nor, as we are, we say we will not shun it'.

Seventeen days and 260 miles later the two armies met at Agincourt. The food the English had taken with them from Harfleur had long since run out and they had been forced to buy, beg and steal provisions along the way. By the time Henry's army lined up to face the French at Agincourt their numbers were somewhere in the region of 8,000–9,000 men.

When they reached the battlefield on 24 October 1415 there was no opportunity to set up their tents. That night it rained heavily, soaking the ground and turning the ploughed field that was to be the site of battle into a quagmire. The English troops sat through the downpour in near-silence, under the threat that if they made a noise they would have their ears cut off. It is a far cry from the scene in the play where Henry goes among his troops, joking with them and offering words of inspiration and comfort to the dejected men.

The Battle of Agincourt took place on 25 October 1415, St Crispin's Day. The English took up position in one line of battle; there simply weren't enough men to do anything else. The vanguard formed the right wing and the rearguard the left, with archers arranged in wedges in between. They were situated on raised ground, a superb defensive position to occupy. The high vantage point also gave them an excellent view of the huge French army as it slowly gathered on the lower ground opposite them.

The French army certainly outnumbered the English but by how much is debated. 'Of fighting men they have full threescore thousand [60,000]' is a wild exaggeration on Shakespeare's part. Certainly there were a greater proportion of men-at-arms on the French side and they would have made an impressive and daunting sight in their armour. From their distant vantage point, the English may have also mistaken the huge numbers of serving men bringing up spare horses from the rear for fighting men. The English might have had the impression that the French army numbered up to 24,000, while the French might have thought they were fighting, at most, 11,000 English (including non-combatants they had mistaken for fighting men). The English were certainly outnumbered, but not by 5:1 as claimed in *Henry V*. There were probably closer to two French fighting men to every Englishman.

Seeing a relatively small number of armoured men on the English side, and underestimating the danger of the archers, the French felt assured that victory was theirs. They spent the eve of battle laughing and making bets on how many English they would capture. Shakespeare shows the French bored with waiting and keen to get on with what they were certain would be an easy victory:

> Lewis the Dauphin: Will it never be day? I will trot to-morrow a mile, and my way shall be paved with English faces.
> Constable of France: I will not say so, for fear I should be faced out of my way: but I would it were morning; for I would fain be about the ears of the English.
> Rambures: Who will go to hazard with me for twenty prisoners?

The mood was very different on the other side of the battlefield. To buoy up his men King Henry rode up and down along the front line giving encouraging speeches. This was an important part of the sovereign's role in war. Even Elizabeth I, not expected to ride into battle personally, still encouraged her troops with stirring words before they embarked for war.* These speeches were a gift for the playwright.

Shakespeare's St Crispin's Day speech is inspiring stuff but not a verbatim record of what was actually said. Even if Henry had uttered the most impassioned speech, few of his men would have been able to hear a word. But it wasn't just about what was said. Seeing the King in his shining armour would have been an encouraging sight – 'that every wretch, pining and pale before, / Beholding him, plucks comfort from his looks'. It was also an attempt to lure the French into battle when they saw their glittering prize within their sights.

Once the speeches are concluded and the French herald, who had again come to try and negotiate peace, is dismissed, Henry gives command of the vanguard to the Duke of York and the battle is under way. In reality there were several hours of waiting as the two sides faced off but no one moved. The English were still in a very good defensive position, which they were reluctant to

* A notable exception to this rule appears in the *Henry VI* plays: Queen Margaret, wife of King Henry, leads the army when her husband is incapacitated, as she did in real life. Her success in battle came as a great surprise to the opposition as well as the men on her own side. In *Henry VI* Part I, Joan of Arc is shown leading the French army with astonishing success. It was such an incongruous sight that the English believed she must be a witch rather than simply a capable woman.

leave, and the French realised they were better off staying put. The French knew that the longer they waited, the more debilitated the English would become and the stronger their army grew as more troops arrived to swell their numbers. Eventually Henry took the unprecedented decision to leave his defensive position and advance on the French. It seemed the only option open to them was to do or die.

* * *

The archers at Agincourt were asked to advance towards the enemy until they were within striking range, around 200 yards. There they would have stopped and stood, wearing almost no protective gear, to fire arrows towards heavily armoured men on horseback charging towards them. Half a tonne of mounted armoured knight could cover the distance in less than 20 seconds. In that time each of the archers could fire maybe three or four arrows to try to bring down the horse or its rider. Being hit by horse and rider going at full gallop would cause injuries that would be described today as 'severe blunt trauma' – the kind of thing you would see in collisions between cars and pedestrians. Death in these cases is caused by internal bleeding as internal organs and major blood vessels are ruptured.

But the troops followed their orders. A huge shout went up across the English lines. Everyone, including knights weighted down in heavy armour, even the 60-year-old Sir Thomas Erpingham, dismounted and ran towards the enemy alongside King Henry himself. The bravery of these men was incredible.

Henry's plan was a brilliant one. The sudden attack took the French by surprise. There was no time to form up into well-ordered, efficient battle lines. They were disorganised and the boggy terrain slowed their advance. Horses slipped and got stuck in the mud. Heavily armoured French men-at-arms were terrifying on horseback, but relatively ineffective in battle once they had been brought down from their saddles.

Battles in this era followed a general pattern of cavalry charge, followed by melee (close combat) and finally the rout (retreat of one side, who were then chased down by the victors). They were undoubtedly chaotic.* The noise would have been phenomenal. It would have been impossible to shout instructions across the battlefield so trumpets were used to give orders. Banners were held aloft as points for troops to rally around.

In the first stage long-distance weapons were used. Large cannons were effective at sieges but difficult to drag across miles of territory to battlefields. At later battles in the fifteenth century rudimentary hand guns were available, but they could be as dangerous to the operator as the target. At Agincourt bows and arrows predominated. The French had short bows that were less powerful, and crossbows that could be deadly but were slow to fire. The English had thousands of powerful longbows and practised archers on their side – the effects of which were devastating.

A steel-tipped arrow, properly fired from a longbow from 30 metres, could pierce armour. At the siege of

---

* Most deaths occurred in the rout rather than the main battle, often from drowning as men tried to flee across rivers.

Abergavenny in 1182, one knight was pinned to his horse by an arrow that had travelled through the flap of his mail shirt, his mail breeches, his thigh and the wooden saddle, and embedded itself in the flank of his horse.

The longbow was almost two metres high (six foot), and required somewhere in the range of 45–80kg, or 100–175lb, to draw the string. It was tailored to the man who would use it. Too light and the bow could be broken by a strong man; too stiff and heavy, and a weaker man would not be able to use it effectively. It required more than strong arms to manipulate a longbow. Standing sideways, the right arm would pull the string back and the left arm would push the bow forward, so the shoulders and back all contributed to the effort. But to be an effective archer required practice and this had been part of English culture since the time of Edward III, enforced by laws passed in Richard II's reign. Generation after generation of archers had practised and honed their skills over a lifetime and passed on their knowledge to others.

A single arrow might be able to kill a single man, but it was thousands of archers unleashing arrows in successive volleys that could terrify an army. At Agincourt maybe 1,000 arrows rained down on the French every second. Even if archers were placed too far away for their arrows to kill, shower after shower of arrows could blind their opponents, wound their horses and throw the enemy into chaos. The invention of the longbow, such a simple weapon but a huge technological advance, changed the nature of battles and the course of European history. Shakespeare makes no mention of the contribution of the archers in *Henry V*. It was perhaps such an obviously well-known weapon and tactic that he didn't feel he needed to.

At Agincourt, the French were hemmed in by the hail of arrows. The battle quickly descended into melee, fought with hand weapons such as swords and axes. There was nowhere to move as more French troops pressed forward from the rear. The French fighters were so closely packed that there wasn't even room to swing their weapons. As each new line of soldiers came face to face with the English, they were ruthlessly cut to pieces until their bodies piled up in heaps over two metres (six feet) high. The ranks that followed had to climb over their dead and dying comrades to try to reach the English front line.

Crushing injuries would have been common. Severely broken rib cages can stop the chest cavity being able to expand and draw in air. Broken ribs or arrows and blades can puncture the lungs. Life-threatening problems can occur when air enters into the area between the lung and the chest wall; this is known as a collapsed lung, or pneumothorax. The air pushes on the outside of the lung, causing it to collapse, and there will be sharp pain and shortness of breath. If the build-up of pressure against the lung is sufficient, blood pressure will drop and the victim goes into shock, leading to death.

Alternatively a 'sucking wound' may form where air is drawn into the lungs through a hole in the chest. The damage is often severe and the injury is complicated by blood loss. If ruptured vessels bleed into the lungs, the blood can displace air as the lungs fill with liquid. Just the pain from these injuries can be great enough to restrict movement.

But some bodies were found on the battlefield, buried under piles of bodies, dead but without a mark on them – they had suffocated. A simple lack of oxygen under the pile of corpses may have been enough to kill many. The crush from the growing pile of both men and horses may have meant some simply couldn't expand their chest from the weight.

Most men, however, would have died from wounds inflicted by swords and axes. Heavy weapons with sharp points and edges were the order of the day and had changed little over the few centuries preceding Agincourt. Swords were the dominant weapon on the battlefield, with battle-axes, maces and war-hammers taking a lesser role.

Rapiers and daggers were light weapons reserved for personal protection on the civilian streets. Swords used in battle were heavier but varied considerably from huge two-handed weapons to short cavalry blades. Shorter swords had a central rib to give strength and a sharp point making them suitable for thrusting and finding out weak points in plate armour. The larger two-handed sword, sometimes known as the old English 'long sword', was a favourite of Henry VIII in his athletic youth. It was a formidable weapon designed to be used against the heavy armour of the Middle Ages. When Henry VIII wanted to include the long sword in tournaments, he was dissuaded by the French King Francis I on the grounds that there were no gauntlets strong enough to save the hands from the powerful strokes of the sword.*

* Alfred Hutton, writing about swordplay in the nineteenth century, advised that actors using long swords in plays should stick very rigidly to choreographed 'sets' rather than 'loose' play as the weapons were so dangerous.

The damage that could be done by a heavy sword was incredible – 'all those legs and arms and heads, chopped off in battle' (*Henry V*). Macbeth during a battle 'unseam'd' his opponent 'from the nave to the chaps'. But, on the whole, surprisingly little has been reported about the details of wounds and injuries sustained in battle. Chroniclers were obviously more interested in the bigger picture, the victories or defeats, rather than the bloody details. Shakespeare has dozens of characters 'slain' but the details of exactly how are few and far between. In *Henry V* the King inspires his men by talking of the wounds they will receive: 'he will strip his sleeve and show his scars, [...] he'll remember with advantages / What feats he did that day.' Battle scars are seen as a source of pride and bravery, and the more the better, as in *Coriolanus*: 'He had, before this last expedition, twenty-five wounds upon him.' But there are no detailed descriptions of their appearance or how they were inflicted.

Some evidence can be collected from archaeological excavations of battlefields, but this is usually confined to wounds that penetrated to the bones. With heavy cutting weapons there is a characteristic lesion in the bones, with one smooth and one rough edge to the wound. The initial impact slices cleanly through the bone on one edge, but the rebound or removal of the weapon is at a slightly different angle, either deliberately or accidentally, which leaves a rough edge.

An injury that leaves its mark on bone is a serious one, but what is surprising is how much damage the body can sustain and still survive. In 1346 King David II of Scotland was struck in the face by an arrow at the Battle of Neville's Cross and it took two barber-surgeons to extract it, but he survived.

More incredible is how long a wounded soldier can remain pain-free and continue with their considerable physical activities. Henry V, at the Battle of Shrewsbury (when he was still Prince of Wales), was struck by an arrow below his right eye. The arrow penetrated six inches. As depicted by Shakespeare in *Henry IV* Part I, Henry had to be persuaded by his father to leave the battlefield for treatment: 'withdraw thyself; thou bleed'st too much.'

Structures within the brain can secrete endorphins in response to stress. Along with adrenocorticotropic hormone (ACTH), a hormone that activates the adrenal glands, endorphins interact with receptors of certain nerve cells. The result is to change normal sensory awareness, so the pain threshold is raised and emotional responses are altered. This is an innate physiological response to protect mammals from danger and pain. It enables the body to continue to function even in life-threatening events. Individuals may be able to carry out normal activities, even to an astonishing degree, in spite of severe injury, only collapsing when the physical damage and loss of blood means the body can no longer function.

In *Henry IV* Part I Hotspur describes how, fresh from the battlefield, 'when the fight was done, / When I was dry with rage and extreme toil, / Breathless and faint, leaning upon my sword,' he had a conversation with another nobleman until the adrenaline rush abated and 'I then, all smarting with my wounds being cold,' loses his patience with the other man.

Cutting injuries from swords and axes, as well as puncture wounds from arrows, were obviously very serious. The size and site of the injury were crucial in terms of chances of survival. Wounds to the neck are the most dangerous. Cuts to the back of the neck that damage or sever the spinal cord result in the loss of nerve signals to the rest of the body. The lungs no longer receive instructions to breathe and the heartbeat is no longer regulated – death is rapid.

The neck also contains major blood vessels with relatively little natural protection, and would have been difficult to protect further with armour without considerable loss of movement. Severing major blood vessels is also a relatively quick way to kill your enemy in battle.

Cuts to the throat also run the risk of air embolism if the jugular vein is cut, though this is rare. Lord Clifford, wounded to death on the battlefield in *Henry VI* Part III, refers to the mischief produced by the contact of air with the wounded surfaces: 'The air hath got into my deadly wounds, / And much effuse of blood doth make me faint.' If he was referring to air embolism it shows remarkable medical insight on Shakespeare's part. If enough air enters the vein it can travel to the right side of the heart where it effectively forms an air-lock. Blood can no longer be pumped through the heart and death is almost immediate. Perhaps instead Shakespeare was referring to infection from the air, which was commonly believed to be the source of contagion. But Clifford seems to expect death to arrive almost immediately, not the relatively lingering, drawn-out affair of infection and sepsis (see later in this chapter).

The more likely outcome of a severe wounding was bleeding to death. Rapid loss of a third to half of total blood volume is fatal. If, for example, the human body loses 5 per cent of its blood volume per minute, that leads to death in six minutes.

When there is insufficient blood within the circulatory system to deliver oxygen to the tissues of the body, the resulting condition is referred to as 'shock'. After the first few pints have been lost there will be heavy rapid breathing as the body tries to compensate for the lack of oxygen normally being transported around the body by the blood. The heart speeds up for the same reason. But eventually internal adjustments are not enough. The heart needs a minimum blood pressure to be able to keep pumping. As the heart itself isn't receiving enough oxygenated blood to perform normally, the heart slows until it eventually stops.

Depending on the injury, the whole process can last a few short breaths or several hours. If a vessel the size of the carotid artery has been severed, this entire sequence can take less than a minute. The speed of death from puncture wounds, whether from arrows or from swords, depends on where the injury occurs and which organs and blood vessels are damaged. The same is true whether the injury was received on the battlefield or on the streets of Verona during a fight between the Montagues and Capulets.

Many of Shakespeare's characters, wounded in a fight, call for surgeons. These surgeons might be able to stem the loss of blood by applying tourniquets – Cassio is saved when a shirt is used to bind a severe leg wound in *Othello* – or using hot metal to cauterise wounds. Basic stitches could also be used to close up cuts. But sometimes the bleeding is too much. In *Romeo and Juliet* Mercutio knows from the severity of his wound that the surgeon cannot help: 'No, 'tis not so deep as a well, nor so wide as a church-door; but 'tis enough, 'twill serve: ask for me to-morrow, and you shall find me a grave man.'

On the battlefield, metal armour would have offered a great deal of protection against a weapon's point. Padded jerkins worn by foot soldiers would have been less effective. Even skin can offer some resistance. But once the skin is penetrated, there is little or no resistance from internal organs. The knife's momentum will push it through tissue until it is stopped by bone or the hilt of the weapon bumping against the body.

Armour could be fantastically effective in protecting its wearer, but there were still vulnerable points. Over the centuries it became more elaborate; in *Henry V* the French lords brag about the quality and beauty of their armour: 'Tut! I have the best armour of the world.' But in reality it had progressed little from the idea of covering as much of the body as possible in heavy metal plates. Joints that allowed even limited movement were also weak points that could be attacked by swords and other cutting instruments. The head was protected by a helmet but that would have hampered communication. The face was covered with a visor, but there was a balance between protection and visibility.

Blows to the head, even if it is protected by a helmet, can be dangerous. There would not be much in the way of cushioning inside a Middle Ages helmet, and the skull itself is essentially a rigid container. The contents of the skull can be deformed by sudden accelerations from blows, causing ruptures of blood vessels that increase pressure on the contents of the skull (see Chapter 5). Bruising to the brain itself causes inflammation at the site of the injury. If the swelling is severe enough then the brain tissue can be compressed to an extent that damage or haemorrhage occurs.

In summary, 'there are few die well that die in battle' (*Henry V*).

★ ★ ★

Modern film productions of Shakespeare's plays can reproduce impressive battle scenes. Laurence Olivier, resplendent in glittering armour, facing the French running at full gallop towards him, banners flowing in the breeze, is an iconic moment in cinema history, but the Bard could not hope to achieve such things onstage. The text skips over the main fighting and proceeds directly from Henry's orders to 'move away' to the latter stages of the battle, after the melee.

As the Battle of Agincourt progressed the French attack weakened. There was enough of a respite for the English to begin searching the piles of bodies for French survivors trapped under the heaps of dead and wounded. Shakespeare shows Pistol, severely hampered by a language barrier, trying to negotiate a ransom from a French soldier he has captured. If he doesn't receive enough money he promises to 'fetch thy rim out at thy throat / In drops of crimson blood.' The scene might be played for comic effect but it portrays the grim reality of how a captive might expect to be treated. Any men of importance were taken prisoner; the common soldier, who would bring little or no money from ransom, had his throat cut.

At Agincourt the prisoners were taken from the battlefield to houses in the small village of Maisoncelle, where they were guarded. Then something happened that caught Henry by surprise. In Shakespeare's version of events, he sees that 'The French have reinforced their scatter'd men' and he immediately orders all the French prisoners to be killed.

Many theories have been offered as to what made Henry take such drastic steps. It might have been an attack at the rear of his army, where the prisoners were being held and could be released to retaliate against the English. It might have been the sight of the French unfurling the Oriflamme, the red war-banner that signified that no quarter would be given. Whatever it was, Henry must have seen the threat as very real and very dangerous. There can be no other explanation as to why he would suddenly order the wholesale slaughter of unarmed men.

Shakespeare justifies Henry's actions by showing the aftermath of a French attack on the English baggage train.

> 'Tis certain there's not a boy left alive; and the cowardly rascals that ran from the battle ha' done this slaughter: besides, they have burned and carried away all that was in the king's tent; wherefore the king, most worthily, hath caused every soldier to cut his prisoner's throat. O, 'tis a gallant king!

Whatever the real reason, it is an astonishing thing to have done. This may have been a time long before war crimes and UN treaties, but there were still accepted rules of engagement and codes of chivalry and honour that armies and their leaders were expected to follow. Henry's actions to kill unarmed, Christian prisoners went against every code of morality and accepted behaviour of the time. Even his own men seemed reluctant to carry out his orders and he had to reissue them under threat of execution to any who disobeyed him. Many would have been reluctant to carry out the orders on moral grounds. Others may have cared more about their financial loss from ransoms.

In the end a group of Henry's soldiers were ordered to go from prisoner to prisoner cutting throats or stabbing them. Perhaps owing to lack of time, some prisoners were left in the houses where they were being held and the buildings were torched (for the agonies of burning to death see Chapter 4). It was a shocking event even at the time, though no contemporary French chronicler criticised Henry's actions.

The orders were carried out, but whatever threat Henry thought he was under failed to materialise. The French were thoroughly defeated. In the play, the line immediately after Henry's repeated order to kill all the prisoners, a French herald arrives with the news that Henry has won the day and asks leave 'That we may wander o'er this bloody field / To look our dead, and then to bury them'.

★ ★ ★

Determining the number that actually died in battle is difficult. Chroniclers at the time were keen to exaggerate the losses on the enemy's side while simultaneously lowering the number of casualties on their own, all for the glorification of the victor. Attempts were made to 'sort our nobles from our common men' (*Henry V*) but the bodies in their blood-soaked, mud-covered and mangled state could not always be recognised; 'So do our vulgar drench their peasant limbs / In blood of princes'. Looters might already have stolen anything worth taking, including armour, weapons and clothes, that might have helped identify individuals. It would be difficult even to sort the bodies into those of the enemy and those of their fellow countrymen.

Names of notable and important men were recorded but the identities of common archers and foot soldiers who perished would not have been noted down, merely counted. It would have been a gruesome task picking your way across a battlefield in search of potential survivors and making a tally of the dead. If the blood and gore wasn't enough there were further horrors to be seen in the immediate aftermath of battle, with arms and legs frozen in awkward poses as rigor mortis set in.

Perhaps Shakespeare is describing a 'cadaveric spasm' when he has the character Montjoy describe the dead 'fetlock deep in gore and with wild rage / Yerk out their armed heels at their dead masters'. This is a rare and virtually instantaneous form of rigor mortis that develops at the time of death. It was said to be particularly common on the battlefield among soldiers slain in combat. Rigor mortis usually begins to appear a few hours after death, but the exertion of battle could mean people quickly used up any reserves of ATP in their muscles and rigor mortis would set in much sooner. The body would be held rigid in the position in which it fell at the time of death.

The death toll at Agincourt was high even by the standards of the day. One estimate of the French losses was indicated by the five grave-pits that were dug, each containing 1,200 ordinary soldiers. In total the French lost 12,000–13,000 people including three dukes, five counts, more than 90 barons and almost 2,000 knights – higher even than Shakespeare's total. Much of the French nobility was obliterated in one afternoon.

The number of English dead was much lower, though not as low as the 25 claimed in Shakespeare's version of events. In total probably less than 1,000 English lost their lives at Agincourt, still a remarkably small number given the odds stacked against them. Among the English dead was Edward of Norwich, Duke of York, the highest-ranking English casualty. According to Shakespeare, York died defending the Earl of Suffolk (who also died), but owing to the chaos of battle it is not clear if he is correct. One account said that the 42-year-old duke rushed forward to protect King Henry when he came under attack. He saved the King's life but lost his own. Other accounts say he was suffocated in the press of men. It has also been suggested his large size, poor health and heavy armour, combined with the physical demands of battle, proved fatal.

The bodies of both the Duke of York and the Earl of Suffolk received special treatment. They were recovered from the battlefield and boiled down to the bones in a huge cauldron brought from England specifically for that purpose. Their bones were gathered together and returned for burial in England. Such consideration was not given to any of the other dead. Even though the number of casualties had been low on the English side, Henry did not feel he had time to remain at Agincourt to bury the dead. Contrary to his promises in the play to bury 'The dead with charity enclosed in clay', in reality the bodies of the ordinary English soldiers were collected up and placed in barns and houses, which were then burned. This may not be as callous as it appears. Henry was keen to get his troops back to England and the threat of further disease and infection from the rotting corpses on an already debilitated army could have been devastating. Henry may simply have been saving time and taking wise precautions against further contagion.

* * *

Apart from the obvious immediate deaths during battle, many men would have died later when wounds became infected and no medicines were available that could kill the bacteria that multiplied in their bodies. The source of infection could be almost anything: the weapons that caused the wounds; the clothing that was dragged inwards into the body; the victim's own skin; or the unsanitary conditions the soldiers lived in. Military surgeons, unaware of germ theory, would not have scrubbed their hands or instruments. Attempts to remove arrows and other fragments from wounds would have introduced even more bacteria.

In the past, the sign of pus oozing from a wound was seen as a sign, not of infection, but that the corruption was being drawn out of the body. Consequently wounds were left open, or 'tented' with material, flax or lint, stuffed into the gap to drain the pus, a practice Shakespeare was well aware of – 'Well might they fester 'gainst ingratitude, / And tent themselves with death' (*Coriolanus*).

Infection causes the body to rapidly produce white blood cells to combat the multiplying bacteria. If the body can contain the infection and its own immune response eliminates the invading bacteria, all may be well. Serious infection, lack of antibiotics and a body already debilitated by the injuries sustained in battle can mean it is less able to fend off the attack from bacteria and other microbes.

Sepsis is the body's response to massive bacterial infection. It follows a predictable pattern. First, there is fever, as the localised increases in body temperature that try to combat bacteria multiplying spread over the whole body. The pulse rate increases and there can be breathing difficulties. An inability to swallow and clear the throat means saliva and other mucus collects in the throat, leading to the gurgling crackle referred to as the 'death rattle'; this is when soldiers in the Middle Ages, or any time before antibiotics and other surgical interventions were available, were likely to die.

If the victim doesn't choke to death, there is worse to come. The blood can start to pool in parts of the body, meaning it is no longer any use in circulating oxygen. Blood clots form and fluid begins to build up around the lungs. Similar processes soon occur around the liver and kidneys. The decreased blood flow to the kidneys prevents proper filtering and results in poor urinary output. This gradually worsens to uremia – the build-up of poisonous products in the blood.

The microbes are also secreting toxic substances and the body's response to these toxins often causes more damage. Chemicals released to counteract them damage cells, including the blood cells, blood vessels and organs. Damage to the blood and vessels means less oxygen is being brought to the tissues, which also become less capable of extracting the oxygen. The result is septic shock. Vital organs fail one after the other.

★ ★ ★

Shakespeare wrote of Henry V's triumphant return to England, but he didn't mention those that died along the way. And Agincourt was not quite the decisive victory Henry might have hoped for, or that Shakespeare depicted on the stage. In 1417 Henry was back in France with another army to continue his campaign to gain lands. In the remaining five years of his life he would spend only five months in England.

The Treaty of Troyes, signed in 1420, promised Henry would inherit the French crown upon the death of Charles VI. But Henry never was King of France, dying in 1422 in miserable conditions while still fighting the French. Charles outlived Henry by less than two months. And although Henry V's infant son Henry VI was crowned King of France (the only English monarch of both countries), his French territories had been lost by the time he reached his majority. The troubled reign of Henry V's only child was the subject of Shakespeare's next three history plays.

CHAPTER SEVEN

A Plague O'both Your Houses!

How foul it is; what rank diseases grow

Henry IV Part II, Act 3, Scene 1

Disease is ever present in our world, but in Shakespeare's time, malnutrition, poor sanitation and ineffective remedies all contributed to much higher death rates. Personal health was understandably something of a preoccupation for Elizabethans and Shakespeare seems to have been more preoccupied than his fellow playwrights. Disease, in some form or other, is mentioned in every Shakespeare play. Mostly it is casual asides, incidental dialogue or a minor event on the sidelines of the main action of the play. A large proportion of Shakespeare's huge stock of insults are disease-based. If a character disliked someone they might literally wish them ill – 'make him / By inchmeal a disease!' (*The Tempest*). In cases of real hatred, particularly nasty diseases might be selected for an especially malicious curse – 'Boils and plagues / Plaster you o'er, that you may be abhorr'd / Further than seen' (*Coriolanus*). But in several plays the role of disease is much more developed.

* * *

The diseases that people in Renaissance England lived and died with were very different from those that worry us today. Bacterial infections such as typhoid, syphilis and scarlet fever, that once killed millions, have been brought under control by the discovery of antibiotics in the twentieth century. Viral infections, such as measles and smallpox, which destroyed countless lives, have been tamed or eliminated completely by vaccinations.

In Shakespeare's day, old familiar diseases, such as tuberculosis, influenza and leprosy, fought for victims alongside virulent newcomers such as syphilis and typhus. It was a crowded and competitive field and not all infectious agents survived. New diseases generally outcompeted old diseases for victims.

The 'English Sweating Sickness', for example, was only a temporary visitor. It appeared in 1485 without explanation and disappeared just as mysteriously in 1551. Cases were confined to England and English-held regions of France; even foreigners living in these areas appear to have been completely unaffected. The symptoms of high fever and profuse sweating appeared overnight or in the morning and continued for 24 hours, after which patients were either on their way to recovery or dead. Mortality rates were around 30 per cent. For reasons unknown, sweating sickness caused particularly high mortality among the upper classes and it therefore received a considerable amount of attention at the time. In *Measure for Measure* Mistress Overdone, a brothel owner, complains that the sweat, among many other causes of death, is having a detrimental effect on her business: 'Thus, what with the war, what with the sweat, what with the gallows and what with poverty, I am custom-shrunk.' The fact that the play was written in 1604, long after the disease had disappeared, and is set in Vienna, where the disease never occurred, shows how much impact it had made on the public consciousness. Though many theories have been put forward, including scarlet fever, influenza and hantaviruses, there is no completely satisfactory explanation of the disease.

One ancient disease, leprosy, saw a notable decline in cases during the fourteenth and fifteenth centuries that coincided with an increase in tuberculosis (TB). As town and city populations swelled, transmission of tuberculosis increased and killed off its victims more quickly. Anyone already debilitated by leprosy was likely to succumb more rapidly to an infection of the TB bacteria.

That leprosy was an unpleasant disease and lepers were generally feared or despised is demonstrated by Timon's reference to leprosy in vitriolic verbal attacks on the people of Athens. Meanwhile the lethal nature of TB, or consumption as it was known at the time, is turned into a joke in *Much Ado About Nothing*. Beatrice and Benedict, despite being at daggers drawn for most of the play, are tricked into falling in love. To show Beatrice has lost none of her fighting spirit she tells Benedict she only agreed to marry him because 'I yield upon great persuasion; and partly to save your life, for I was told you were in a consumption.'

Malaria (literally 'bad air' because the disease was associated with swamps and other damp areas) was known as 'ague' in England up until the eighteenth century. It is mentioned in eight of Shakespeare's plays, most notably in *The Tempest,* which is set on a magical malarial island. Prospero, a powerful sorcerer, lives on the island with his daughter Miranda, a spirit called Ariel and the monstrous Caliban, who has been enslaved to do Prospero's bidding. More people arrive on the island because of a shipwreck.

Caliban curses his master by wishing malaria on him: 'All the infections that the sun sucks up / From bogs, fens, flats, on Prosper fall'. The curse shows an understanding of the environmental conditions that led to malarial outbreaks. They mostly occurred in late summer and affected people living close to brackish water. Malaria is transmitted by mosquitoes, insects that breed in stagnant water. The disease itself is caused by a plasmodium, a type of single-cell parasite that develops inside the mosquito and is injected into a vertebrate host (such as a human) during a blood meal. The parasites then multiply in the host's liver cells before entering the bloodstream to infect red blood cells. This destroys the host's red blood cells, resulting in the malaria symptoms.

In the play one of the castaways, the butler Stephano, meets Caliban when he is in a drunken stupor. He mistakes his alcohol-induced trembling and delirium for an attack of malaria: 'hath got, as I take it, an ague [...] he's in a fit now and does not talk after the wisest'. Stephano tries to help Caliban through the worst of the attack: 'He shall taste of my bottle: if he have never drunk wine afore it will go near to remove his fit'. The wine was expected to 'shake your shaking'. Alcohol may have made the patient feel better but it wouldn't treat the infection itself. Effective antimalarial treatments, such as cinchona powder, were not available in England until 1660.* Malarial outbreaks in England declined as cinchona remedies became more widely available. The drainage of marshes and fens for agricultural use also meant there were fewer places for the mosquito to breed.†

Once contracted, malaria can persist for years with occasional relapses caused by the blood stages of the plasmodium's life-cycle. Fevers described as tertian or quartian referred to the regular cycle of fevers, and match well with the symptoms from infections of malarial species *P. vivax* and *P. falciparum*. The reference to a tertian quotidian in *Henry V* suggests it might have been malaria that killed Sir John Falstaff (see Chapter 3).

In Shakespeare's day infection and illness would have been common and accepted as part of life, but a handful of diseases were particularly dreaded because of their high mortality rate or social stigma. Smallpox, caused by a virus, killed a large proportion of those it infected and left disfiguring scars on those it didn't kill. Syphilis, a bacterial infection known as the pox, was associated with illicit sexual behaviour and could affect

---

* The bark of several species of cinchona tree contain quinine, the first known effective drug for the treatment of malaria.

† Other factors also contributed to a reduction in malaria cases, but it didn't disappear from Europe until the twentieth century and the invention of DDT.

a person's health as well as their social standing for years. Above all others, plague terrified communities because of its sudden appearance and swift and devastating mortality rate.

For a disease that was so pervasive in the Elizabethan and Jacobean eras, it is surprising that there is no plague play, or certainly none that has survived, and references to the disease in existing plays are minimal considering the impact it must have made on people at the time. The references that do exist, are usually insults or remarks to express disgust. Shakespeare is liberal with using plague as a general curse, usually against people, but also against anything annoying, from drums (*All's Well That Ends Well*), to the weather (*The Tempest*) and even pickle-herrings (*Twelfth Night*). It either shows an absolute and serious hatred for recipients of the oath, or gallows humour of the blackest kind. Laughing at adversity may be one way Elizabethan audiences coped with grim reality. The pestilence held the population in constant dread. No playwright depicted plague in any realistic way or detailed its awful effects. It is almost as though the topic were too terrifying to mention or show onstage. The theatre was an escape from everyday worries and audiences didn't need reminders of the reality of the terrible pestilence.

★ ★ ★

Plague had been endemic in England since the Black Death in the fourteenth century, with occasional devastating major outbreaks. In the seventeenth century scarcely a year went by without at least a few deaths from plague until the last major outbreak in 1665. Plague shaped Shakespeare's life. It closed theatres and killed his fellow actors; it even turned the playwright into a poet when he was forced to find alternative sources of income.

Plague outbreaks were more than an inconvenient interruption. It struck without warning, was agonising and unpleasant to suffer, and there was a high chance it would kill you, so people were justifiably frightened. In *Romeo and Juliet*, when Mercutio is fatally wounded, he spits out the most terrible curse he can think of on those responsible: 'A plague o'both your houses!'

The plague bacteria, *Yersinia pestis* (commonly known as bubonic plague, after the buboes or swellings that are characteristic of the disease), can infect over 200 different species, not just humans. The natural hosts of the disease are burrowing rodents that have lived with plague for thousands of years and have therefore developed resistance. But when the plague jumps from rodents to humans it becomes particularly virulent. Sixteenth-century Europe was an environment with many opportunities for plague to cross over from rodents to humans.

It was, however, an exceptional period. Over the course of human and plague history our exposure to the disease has been too infrequent for our species to adapt and evolve to defend ourselves. There is evidence that some people who survived plague acquired some kind of immunity, but this was not passed on to future generations. Recurring outbreaks would be expected to affect all ages, but in the sixteenth and seventeenth centuries it tended to be mostly children that were infected. Maybe Shakespeare's early exposure to the bacteria as a child helped him survive the numerous outbreaks he was exposed to as an adult.*

Plague is transmitted not by the rodents themselves but by their fleas. All fleas bite, but not all flea bites are the same. Some are particularly good at transmitting plague, especially *Xenopsylla cheopis*, the oriental rat flea. This is because the bacteria can form a blockage between its oesophagus and mid-gut. This blockage is a mixture of the last blood meal and sticky plague bacteria. This is bad news, both for the flea and for the animal it bites. The blockage means the flea will eventually starve to death, but not until a few days of vigorous biting have passed. When it tries to take in a new blood meal this cannot pass into the gut, so, instead, it is vomited back into the animal that has been bitten, along with the plague bacteria.†

Once inside the body *Yersinia pestis* moves through the blood to the lymph nodes where it replicates. The bacteria carefully conceal their activities from the immune system of their new host. They are able to hide by forming a protective envelope that resists the normal process of elimination, phagocytosis (being swallowed up and digested by immune cells). *Yersinia pestis* also prevents immune cells from signalling distress that would bring more immune cells to fight off the infection. So at

---

* Plague eventually died out in Europe, not owing to a natural build-up of resistance, but because there is no native species of burrowing rodent that can allow the plague bacteria to survive the cold winters.

† Not all flea species form blockages and it isn't necessary for plague to be transmitted from one host to another, but it most certainly speeds up the process.

this early stage of infection there is no inflammation, or tell-tale sign that the victim is infected, and the bacteria continue to multiply unhindered.

At some point the growing ball of bacteria will become so large that it bursts the immune cells and spills into the bloodstream. A sudden influx of bacteria into the blood system would normally cause septic shock that would kill the host before the infection could be passed on (fleas don't bite dead animals). So the plague bacteria modify the body's normal response. A molecule on the surface of the majority of disease-causing Gram-negative bacteria, known as Lipid A, is responsible for most of the bacteria's toxicity. At normal body temperature the plague bacteria produce a modified version of Lipid A that is much less toxic. This allows the bacteria time to evade the immune system and for a flea to feed on the infected blood and transfer the bacteria to a new host.

A person can feel quite well for up to six days following the initial infected bite. Then they will be suddenly struck down when the infection reaches a critical point. By the time symptoms appear, their tissues are already riddled with bacteria. To Elizabethans it seemed as if the plague appeared out of nowhere without warning – 'Even so quickly may one catch the plague?' (*Twelfth Night*).

The well-known symptoms, buboes in the groin, armpit and neck, are the lymph nodes swollen with bacteria. The buboes become huge and exquisitely painful: 'a boil, / A plague-sore, an embossed carbuncle', as Shakespeare puts it.* Multiple buboes can arise from

* This could also be interpreted as a reference to a syphilitic sore, which we will discuss later in the chapter.

multiple flea bites, producing 'plague tokens', 'God's tokens' or 'death-tokens'. In *Love's Labours Lost*, a play full of verbal tricks and cleverness, Biron plays on the multiple use of the word token (love/death-token and Lord's/God's token) when he teases the ladies that they have fallen in love with their suitors after receiving their gifts: 'They have the plague, and caught it of your eyes; / These lords are visited; you are not free, / For the Lord's tokens on you do I see.'

Sometimes these buboes turn black and rotten. Sometimes they slough off to leave behind rotten tissue, exposed muscle and occasionally even bone. Alternatively the buboes ripen and discharge large quantities of pus; undoubtedly an unpleasant experience, but in the past this scenario offered the best chance of recovery. Nevertheless, once contracted, the prognosis was poor – 'He is so plaguy proud that the death-tokens of it / Cry "No recovery"' (*Troilus and Cressida*). Between 40 and 60 per cent of those affected died. This is a terrifyingly high proportion considering one of the most dreaded diseases of modern times, smallpox, killed 20–30 per cent of those infected.

Although *Yersinia pestis* has evolved brilliant methods for evading the immune system and facilitating its transmission between hosts, those hosts would normally be rats. Rat fleas prefer to feed on rats, not humans (humans have their own variety of fleas). Fleas are fairly homey creatures, and tend to stick with their dinner once they have found it. They amble around with the rat as it goes about its business, biting their host three or four times a week. Only when the rat dies will the flea reluctantly leave to find a new host. *Yersinia pestis* has to kill the rat host eventually in order to spread. During a major outbreak, when lots of rats die, the fleas start to look for alternative accommodation, and humans are a convenient substitute even if we don't taste quite as good.

In the late sixteenth and early seventeenth centuries humans and rats lived in particularly close proximity. Housing was rarely of a quality that would keep out rodents. They also easily became dependent on human detritus left on the London streets. The clustering of plague cases in the poorer, more densely populated and dirtier parts of town illustrates the problem.

Rats are also not very adventurous creatures, rarely travelling more than a few streets away from where they were born. The spread of plague over vast distances is the result of human activity. Trade provided the methods of transport that brought these hosts and their infected fleas from one city to the next in wagons, or from port to port in ships. Shakespeare describes the potential problem of large numbers of people travelling regularly and over long distances in the Renaissance era. He has a carrier stop off at an inn during his travels and complain of the poor quality of accommodation, caused by so many people passing through: 'I think this be the most villainous house in all / London road for fleas: I am stung like a tench' (*Henry IV* Part I).

Fleas were far more common than they are today and the poor state of personal hygiene would only have attracted more of them. People often wore the same clothes and slept in the same bed-linen for long periods of time without washing them. In *The Merry Wives of Windsor* the Fords are a prosperous couple with money to spend on clothes and laundry, but their affluence does not mean their lives are flea-free. In the play Ford is searching a laundry basket for the man he thinks is having an affair with his wife, but she warns him, 'If you find a man there, he shall die a flea's death.'

Human fleas, *Pulex irritans,* probably contributed to the spread of pestilence as well.* They do not form the same gut blockage that makes rat fleas such a good transmitter of plague bacteria. But frequent biting still makes *Pulex* a likely means of transmission between individual members of a household who were often living in cramped homes and sharing beds.

If bubonic plague wasn't bad enough, there were other forms of the disease that were more terrifying still. Plague comes in three main varieties: bubonic, septicemic and pneumonic. Bubonic plague is by far the most common form. Septicemic plague is the rarest, infecting the blood rather than the lymph nodes. It causes a breakdown in blood-clotting processes. Lots of tiny clots form throughout the body, depleting clotting resources, so that uncontrolled bleeding then occurs because new clots cannot form. Untreated it is almost always fatal.

In some circumstances (the details are still not certain) *Yersinia pestis* can move to the lungs to produce the third form of plague, pneumonic, the most frightening of all. Infection in the lungs produces hacking, bloody coughing that can transmit the bacteria directly between individuals via droplets of blood expelled from the lungs, meaning there is no need of an intermediary flea.

---

* Although this theory is accepted by experts in France and Russia, it remains controversial in the US.

Pneumonic plague kills its victims within three days and virtually no one survives. The Black Death of fourteenth-century Europe was such a devastating plague outbreak because the pneumonic form was present as well as bubonic and septicemic forms. In *Macbeth* Ross describes the miserable life in Scotland under Macbeth's rule: 'good men's lives / Expire before the flowers in their caps, / Dying or ere they sicken', but he could as easily be talking about life during the Black Death.

There were at least five major outbreaks of bubonic plague in London during Shakespeare's lifetime and though these outbreaks didn't reach the devastation of the Black Death, they all had a major impact on the population, particularly in towns and more populated areas. Wealthier Londoners often took Chaucer's advice, written during the Black Death, to 'run fast and run far'. At that time there were few uninfected corners of Europe that you could run to. At least a quarter of Europe's 75 million population died in the mid-fourteenth century.* The plagues of the Renaissance were a different matter. Escaping the city during the fifteenth- and sixteenth-century outbreaks would have significantly improved a person's chances of survival. Shakespeare was fortunate to have a house and family in Stratford that he could retreat to when plague appeared in London.

There was some recognition that plague was contagious, even if the mechanism was far from understood. Some

* At the time King Edward III was on the throne in England, but when Shakespeare collaborated with Thomas Kyd to write a play about the King they made no mention of plague at all. Instead the play focuses on Edward's successes fighting the French and completely ignores the devastation from pestilence.

suspected it was brought to London by foreigners. Others tried to blame outbreaks on an unusual alignment of the planets. The 1593 plague was blamed on the position of Saturn in the night sky 'passing through the uttermost parts of Cancer and the beginning of Leo' as it had done 30 years earlier when there had been another terrible outbreak. Shakespeare was certainly aware of the planetary theory, as in *Timon of Athens* the playwright has Timon urge Alcibiades to take revenge on Athens: 'Be as a planetary plague, when Jove / Will o'er some high-viced city hang his poison / In the sick air'.

The mention of vice in the same passage acknowledges that many saw plague as punishment from God. It was just reward for the licentious living for which city dwellers were renowned. This position was difficult to maintain when priests, expected to visit the sick and dying and therefore especially susceptible to infection, suffered particularly high mortality rates from the disease. What was clear was that when one person died of plague others closely associated with the sick often became ill themselves.

Civic authorities did what they could to manage outbreaks when they occurred, but without a true understanding of the disease and how it was transmitted, they could put up only limited effective defences. Public-health measures during disease outbreaks were patchy at best. Bonfires were lit in the streets in an attempt 'to purge and cleanse the air' and stray dogs were culled. But most efforts were concentrated on trying to contain an outbreak when it occurred. Ships were held at anchor, travellers on foot were held in special hospitals for forty days (*quarantine* derived from quaranta, Italian for 'forty days') before entering the city and playhouses were closed, though churches remained open.

Some plague victims were taken to 'spitals (hospitals) or the 'pest-house' where nursing care was notoriously poor. Once inside you were not expected to re-emerge. Mortality rates in the pest-houses have been estimated at 98 per cent during the 1665 plague outbreak. Many of these measures amounted to simply separating the sick poor from the nervous rich.

If plague was discovered in a home, the house would be sealed up with the sick and the well trapped together inside until the plague had passed. A red cross was painted on the door as a warning to others. But red crosses and locked doors did nothing to prevent rats from moving in and out of houses, and in fact the idea of preventing the spread of plague by containing it may have resulted in more deaths. Cooped up together, transmission between individuals would have been easy. And at least one family died of starvation because they were shut up in their house in London during the 1592–3 outbreak.

The unpleasant and potentially risky task of entering houses looking for signs of plague was given to searchers. These were recruited from the ranks of the most disposable members of the community – elderly women of low social status. They were 'honest, discreet matrons' who lived apart and received four to six pence for each plague body they identified. These women had no medical training and undoubtedly made mistakes, not necessarily through malice but through plain ignorance or fear.

Households locked up and separated from the outside world were a familiar sight and therefore a credible plot device for how important messages could be unfortunately delayed. In *Romeo and Juliet,* plague, or fear of it, prevents Friar John from bringing Romeo the vital information that Juliet will not be killed by the poison she has drunk but merely appear to be dead. Friar John explains:

> the searchers of the town,
> Suspecting that we both were in a house
> Where the infectious pestilence did reign,
> Sealed up the doors, and would not let us forth,
> So that my speed to Mantua there was stayed.

Shakespeare borrowed the plague plot device from earlier versions of the Romeo and Juliet story and it is surprising that such a simple and credible way of diverting, delaying or even killing off characters wasn't used more. Plague may be conspicuous by its absence onstage but many others were writing about it. People made huge profits from the sale of treatises, pamphlets and books on the prevention and cure of plague (23 books were published on the subject between 1486 and 1604). Yet more money was made by those selling remedies and supposed cures.

Contemporary treatments for plague were staggering in their number and diversity. And they were all almost completely useless. That didn't stop them from being sold or bought in great quantities. Exotic ingredients were suggested to those that could afford them. For example, treacle* and gunpowder could be used 'to provoke a sweat'. But the poor were not forgotten and it was suggested that

* Treacle may be a corruption of the word theriac, the name used for a general all purpose remedy that sometimes included treacle as an ingredient in English recipes.

they 'may eat bread and Butter alone, for Butter is not only a preservative against the plague, but against all manner of poisons'. One recommendation was that a live plucked chicken should be applied to the plague sores to draw out the disease. Smoking came highly recommended as the haze of smoke would ward off the foul smells thought to cause disease. Tobacco, recently introduced to Europe and popularised by Sir Walter Raleigh, was therefore seen as a wonder drug. It was a relatively rational approach to plague prevention as it was understood at the time; unfortunately, the disease was not in the air. Smoking didn't stop plague but tobacconists made a killing.

⋆ ⋆ ⋆

Like plague, dysentery was another lethal infection too terrifying to show on the stage. It was a particular scourge of armies on campaign. One early name given to the disease, 'the bloody flux', was certainly descriptive of its symptoms. Dysentery is caused by a number of different bacteria, viruses and parasites, and produces agonising abdominal pains, inflamed bowels and excessive, and bloody, diarrhoea.

Despite depicting many wars in his plays, Shakespeare only referred to dysentery in *Henry V*, and even there it is only alluded to as a sickness. It is likely his audience was perfectly familiar with the disease and didn't need a detailed description. Also, Shakespeare didn't want to dwell on Henry V's near-failures at the siege of Harfleur, but to promote his successes. At Agincourt he wanted to show that an English army, even weakened by disease, was still more than a match for the French.

At Harfleur, while cannonballs and other missiles pounded the city walls, Henry and his men were camped outside. The people trapped inside the walls had control of the waterways around the town and deliberately opened sluices to flood the fields to the north. This severely hampered where Henry could move his troops and the stagnant water in the heat of the summer soon became a prime spot for breeding bacteria and bugs.

The outbreak of dysentery affected not just Henry's men, and those located near the King's tent in particular, but also those inside the town. Henry had to send over 2,000 troops back to England to recover. These men were effectively lost to Henry's army, though few lost their lives. Shakespeare has Henry admit as much to the envoy from the French army: 'My people are with sickness much enfeebled, / My numbers lessened'. Although some contemporary chronicles claim the death rate from disease was very high, there is only direct evidence for 37 deaths and modern historians estimate dysentery killed around 50 people at Harfleur.*

The disease had no respect for rank; it killed the meanest foot soldier and kings alike. The death of Bishop Courtenay during the siege of Harfleur was a considerable worry. Not only had the disease claimed a man of the cloth, but the Bishop had often stayed in the King's quarters. Henry was lucky to escape infection on that occasion but his luck ran out seven years later in 1422.

* Contemporary chroniclers, who were not there in person and wished to promote Henry V's achievements against all the odds, probably exaggerated the devastation caused by disease, or were merely confused by those that were sent home sick.

Again he was on campaign in France. At the siege of Meaux, Henry contracted dysentery but struggled on towards Paris. As the disease progressed he was in such pain that he could not ride his horse and eventually had to be transported by boat along the Seine. By the time he died he was skeletally thin. Shakespeare tactfully ignored Henry's miserable death and ended his play on a high note with the acquisition of the French Princess Catherine as his bride and, by implication, the French throne.

Henry V was not the only king to die of dysentery. King John fell prey to the disease while fighting against his own barons. Though he had hoped to continue his campaign, he fell ill at Newark and was unable to continue. He succumbed to the disease during the night of 18–19 November 1216. Contemporary chroniclers had several alternative theories as to what had killed the monarch, and Shakespeare chose the most colourful when he portrayed the event in *King John*: he laid the blame on a monk who had poisoned him (see Chapter 8).

Dysentery wasn't the only disease that worried the military. Soldiers that survived the 'bloody flux' might have had the opportunity to contract venereal diseases from the prostitutes who often accompanied the troops. When soldiers returned home from campaigns they often took booty home with them, the spoils of war, and they may also have carried syphilis, which spoiled them in a different way. Playwrights may have been reluctant to portray the horrors of plague or dysentery but they were more than happy to satirise syphilis and its sufferers.

* * *

Victims of disease, and particularly disfiguring diseases such as smallpox or scrofula, were usually seen as objects of pity. But syphilitics, with their characteristic facial deformities, were openly mocked. Syphilis, most commonly referred to as 'the pox',* first appeared in Europe in 1494–5 but its spread was rapid.† By the time Shakespeare was writing it was such a common disease that his audiences would have easily understood even very subtle references to it in his plays and poetry.

At the time, the cause of the disease was unknown but many theories were offered up. Some believed the disease had been created during the sexual union of a courtesan and a leper. Others thought it a curse from God for illicit sex. Still others laid the blame on anything from cannibalism to American iguanas or the potato. However, some commentators were more perceptive.

In 1546 Girolamo Fracastoro composed a 1,300-line poem about the pox. Hidden within those lines he postulated the existence of tiny invisible living things that could cause the disease: what today we would call germs. Fracastoro was centuries ahead of his time. Germ theory was not widely accepted until the mid-nineteenth century and the germ that causes syphilis was not identified until 1905 by Erich Hoffmann and Fritz Schaudinn.

What Hoffmann and Schaudinn identified was the spirochete, or bacterium, *Treponema pallidum*. It is a bacterium with a snakelike shape about the size of a red

---

* Not be confused with smallpox, which is a completely different disease.

† Confusion with other skin diseases has caused some doubt about the exact date.

blood cell. The spirochete wriggles its way from one person into another through a break in moist skin or through mucous membranes. The mystery of syphilis has now been made clear, but where it came from is still vague.

The first descriptions of the disease showed it was new, or certainly a much more vigorous and devastating form of an existing disease. When new diseases are introduced into a population they are initially more virulent as there is no previous immunity to defend against them. The symptoms in these initial infections are more exaggerated, and can even differ markedly from the symptoms displayed by patients once the disease has become established in a population.

The first great syphilis epidemic occurred in Naples in 1494–5 during a French invasion. It produced grotesque sores all over the body. The disease swept through the population so dramatically and devastatingly that even lepers protested when syphilitics moved into their neighbourhood. By the end of the sixteenth century milder strains of syphilis had displaced more aggressive forms that killed their host too quickly and they have been with us ever since.*

The first appearance of the disease in Europe came about very shortly after the return of Columbus from the New World on 15 March 1493. For many this is more than mere coincidence. The assumption is that Columbus and his crew brought back the disease from Hispaniola (an island divided between modern-day Haiti and the Dominican Republic). But there is an alternative theory.

---

* Over the 500 years from its first appearance in Europe, syphilis may have been responsible for 10 million deaths in Europe alone.

Some believe the initial epidemic in Naples was caused by several diseases, one of which was syphilis, combining to cause such devastating results. The theory goes that syphilis was already present in Europe in the form of yaws, a subspecies of the same bacteria that causes syphilis, *Treponema pallidum*. Yaws was certainly in existence in Europe long before syphilis arrived, and it would have been classified as leprosy by medieval doctors because of its effects on the skin (it causes deep open sores to appear). It is transmitted by direct contact with the skin of an infected person.

The theory states that syphilis is a mutation of the yaws spirochete into a more dangerous form that is transmitted through mucous membranes by sexual contact. It is suggested that the mutation would have been prompted or exacerbated by the mass movement of human populations, such as happened within the slave trade.

Yaws could be caught at any age and was likely to be an early childhood disease that would help the population build up a tolerance to it (childhood symptoms are often milder than if the same disease infects an adult). But syphilis would have infected people when they were older.

Evidence of syphilis has been found in skeletons from the Dominican Republic dating back to pre-Columbian times. This appeared to end the argument once and for all in favour of the Hispaniola/Columbus theory. That held true, until skeletons were excavated from a medieval monastery known as Blackfriars in Hull, England. The bones were dated to between 1300 and 1420, long before Columbus's return, and appeared to show signs of syphilis. Others claimed the skeletons showed signs of yaws, not syphilis. The debate continues to rage and is only likely to end when a test can be established to distinguish the yaws spirochete from the syphilis one.

Whatever the origins of the disease, the outbreak in Naples in 1494/5 marked a turning point. It quickly developed into a global epidemic. The new, more active, form of *Treponema pallidum* spread astonishingly quickly. It was undoubtedly helped by war, with disbanded soldiers carrying the infection far and wide. By 1498 the disease had reached India. From there it moved to China where it was known as the 'Canton rash'. By 1512 it had arrived in Japan under the name of the 'Chinese ulcer'. As the contagion spread, each country blamed its neighbour. The Russians blamed the Poles, the Poles blamed the Germans, the Germans blamed the Spanish. The French and Italians blamed each other, as did the Christians and Muslims.

By the time Shakespeare was making satirical references to the disease in his plays and poems, the pox was responsible for up to half of all hospital admissions, according to the surgeon William Clowes.* Prostitutes were blamed for incubating the disease but few blamed their clients for the spread of the infection. Falstaff voices the common belief when he tells the prostitute Doll Tearsheet, 'If the cook help to make the gluttony, you help to make the diseases, Doll. We catch of you, Doll, we catch of you' (*Henry IV* Part II). There were thousands of women selling sex in England's capital and other major towns, to eke out a living in difficult economic times. Certain districts became well known for their

* Clowes may have been exaggerating in order to promote his medical work as well as the book he had published, *A Short and Profitable Treatise Touching the Cure of the Disease Called Morbus Gallicus.*

brothels, and the South Bank was home to some of the most notorious brothels or 'stews' as they were known.

The South Bank was also home to Shakespeare's Globe theatre and the site of the Bishop of Winchester's palace. Many of the South Bank brothels were formerly controlled by the Bishop of Winchester, until the law changed and there was a crackdown on prostitution in Henry VIII's time, so prostitutes gained the nickname of Winchester geese. A Winchester goose might also be someone who had contracted syphilis from a prostitute. In *Henry VI* Part I the Duke of Gloucester calls the Bishop of Winchester a 'Winchester goose', which is a much greater insult than it may appear. The same phrase is used in *Troilus and Cressida*, which contains so many references to syphilis that it has become known as the 'pox play'.

The play was inspired by Homer's *Iliad* and Chaucer's *Troilus and Criseyde*.* The two named in the title of the play are lovers brought together by Cressida's lecherous uncle, Pandarus (the name gave rise to the Elizabethan slang 'pander' for a bawd or pimp). If the play's setting of war, and Pandarus's arranging illicit sexual meetings between the lovers, wasn't enough to associate the play with syphilis, there is more.

The lovers are separated when Cressida is handed over to the Greek side in exchange for a Trojan prisoner. Despite swearing to be faithful to Troilus, Cressida is wooed by Diomedes, showing women's perceived

* In the *Iliad*, Troilus is a brave soldier without the taint of sexually transmitted disease about him, but then it was written centuries before the disease appeared in Europe. Chaucer added the romance with Cressida and Shakespeare then embellished it with disease.

unfaithfulness. Another Greek character, Thersites, plays the part of the fool hurling insults at everyone, usually laced with references to disease. One particular insult, directed at Patroclus, lists the symptoms of syphilis:

> Why, his masculine whore. Now, the rotten diseases of the south, the guts-griping, ruptures, catarrhs, loads o' gravel i' the back, lethargies, cold palsies, raw eyes, dirt-rotten livers, wheezing lungs, bladders full of imposthume, sciaticas, limekilns i' the palm, incurable bone-ache …

The list is a long one, and such detailed knowledge of the disease might not be expected from a writer with no medical training, but Shakespeare potentially had access to a number of other sources (see Chapter 2). Apart from consulting doctors, there were many treatises and commentaries written about the disease that he could have learned from. For example, Phillip Stubbes wrote in his 1583 *Anatomie of Abuses*:

> On whoredom […] besides that it bringeth everlasting damnation to all that live therein […] it also bringeth these inconveniences, with many more, videlicet [viz], it dimmeth the sight, it impaireth the hearing, it infirmeth the sinews, it weakeneth the joints, it exhausteth the marrow, consumeth the moisture and supplement of the body, it revileth the face, appalleth the countenance, it dulleth the spirits, it hurteth the memory, it weakeneth the whole body, it bringeth consumption, it causeth ulceration, scab, scurf, blaine, blotch, pockes and biles, it maketh hoare haires, bald pates, induceth old age, and, in fine, bringeth death before nature urge it, malady enforce it, or age constrain it.

Stubbes may have conflated syphilis with more than one other disease, but you get the general idea.

The list could be even longer if all the possible symptoms of the disease are listed. As John Stokes, a twentieth-century specialist in syphilis, put it, 'Almost anything might be expected at any time.' The symptoms mimic so many other diseases in the later stages that people were often misdiagnosed. Syphilis became known as 'the great imitator'. The crucial initial symptom of diagnosis, a chancre on the genitals, may have healed years before and long been forgotten.

The initial often painless chancre, or sore, usually appears on the genitals around three weeks after infection; it is followed by fever, rash and a feeling of malaise. This stage is known as primary syphilis. The sore will disappear after about six weeks but the spirochete has already moved into the rest of the body, through the bloodstream, where it settles down for a prolonged stay. The body's immune system may put up a valiant fight, destroying millions of bacteria, but a few of the spirochetes will evade the onslaught and go on to reproduce. All tissues are likely to suffer over the following years and decades, with nerves and blood vessels being particularly susceptible to attack.

After the initial stage the symptoms subside for an indeterminate length of time, from a few months to many years, but averaging seven years. During this period many would have thought their initial treatment had been successful. But the spirochete was still there, multiplying and causing a slow progressive inflammatory reaction. This symptom-free time was most dangerous, as syphilitics could still pass on the infection to others. The disease is very infectious for the first two years, less so after that and rarely contagious after five years.

After the period of latency, the disease re-emerges as an all-over rash of pustules, much like chicken pox. References to 'lazar-like' eruptions of the skin in Shakespeare's plays may show his knowledge of this later development of symptoms. This stage is known as secondary syphilis.

The disease will disappear from view again but leave the victim feeling dreadfully unwell. Patients often went from doctor to doctor to describe mysterious fevers and aches and pains with no apparent cause. This is the third stage of the disease, tertiary syphilis, when the inner structures of the body are attacked. The spirochete can create symptoms that look like anything but syphilis. Before a clinical test was developed, people were diagnosed with anything from rheumatism to gout, eczema, epilepsy, jaundice, schizophrenia or just plain 'nerves'. In the final stages, tertiary syphilis develops into a hideously disfiguring and agonising disease.

One of the characteristic disfigurements of tertiary syphilis is a collapsed or 'saddle' nose, caused when the cartilage of the nose is eaten away.* Sir William Davenant, poet, playwright, godson and (he claimed) actual son of William Shakespeare, 'got a terrible clap of a Black handsome wench that lay in Ale-yard, Westminster [...] which cost him his Nose.'

In the past corks were stuffed into nostrils to give a more normal appearance to the nose. Quills were then inserted to allow the individual to breathe. Some

---

* This is rarely seen today as syphilis can be treated effectively early on with a course of antibiotics.

Portrait of Sir William Davenant by John Greenhill.

attempted to cover up the deformity by wearing a false nose made of copper. In *Troilus and Cressida*, Cressida, resistant to her uncle's attempts to persuade her to like Troilus, commended him 'for a copper nose'.

Finally death would arrive after years or decades of poor health. This could be a slow protracted death, writhing in a bed raving about God (from the spirochete attacking the brain), or a very sudden event when an aortic aneurysm, caused by the spirochete attacking the blood vessel's walls, suddenly gave way.

Some took precautions against infection by wearing 'venus gloves' or condoms (in *Troilus and Cressida* apparently Menelaus's ex-wife swears by them). It is doubtful these early condoms were as effective as modern versions in preventing transmission, but they were considerably better than other preventative measures that were taken. Some who engaged in risky activity took the precaution of wrapping their endangered organ in a piece of cloth soaked in wine, shavings of Guaiac (tree resin), flakes of copper, mercury compounds, gentian root, red coral, ash of ivory and burnt horn of deer. Stewed prunes were thought to be an effective cure and were often served in brothels. 'Hard pissing' was thought to flush out the disease and so brothel owners often placed two chamber pots under each bed for their clients.* At a time when all diseases were thought to be due to an imbalance of humours, removing the 'corruption' by bleeding, vomiting, sweating, defecation or urination was thought a reasonable treatment, but they were wrong.

A burning sensation during urination is a typical symptom of venereal disease, but of gonorrhoea rather than syphilis. In Elizabethan England it was often conflated with syphilis under the general term 'pox', though it was likely that both infections were often contracted simultaneously. Some thought the burning sensations of gonorrhoea were just the prelude to syphilis and they were different stages of the same disease. It also gave an opportunity for Shakespeare to add in a few knowing jokes and an alternative interpretation to phrases such as 'I burn with thy desire' (*Henry VI* Part I) and 'Love's fire took heat perpetual' (Sonnet 154).

If preventative measures failed, and a chancre did appear, early treatment might be sought. But there was a problem. The pre-eminent medical authority of the time was Galen, a second-century Greek physician who

---

* Urinating was also erroneously thought to act as a contraceptive.

ministered to gladiators and emperors in the Roman empire. Galen had no treatment for pox as the disease was unknown in his lifetime, so sixteenth-century medical men improvised. One treatment involved covering the sore with a spider's web (an adaptation of a treatment also used to heal cuts). A more drastic remedy would be to attempt to remove the sore. If that didn't work, binding the base of the penis was thought to prevent the infection spreading to the rest of the body. None of these treatments would have helped and as the disease progressed, symptoms got worse.

New and virulent diseases demanded new and stronger treatments. One of the most common treatments for pox was mercury, leading to the phrase 'A night with Venus, a lifetime with Mercury'. Mercury had long been used in Arabic medicine to treat skin disorders such as leprosy, and it was easy to make the link with the skin lesions that appeared with syphilis infections.

Mercury, or a mercury compound, could be applied directly to the affected area, delivered by mouth or used as a fumigant. Patients would be seated in a tent-like structure with their head emerging from the top. Under their seat would be a heated pan of mercury. In *Henry V*, when Pistol is speaking about the fate of his 'Doll',* he talks of her treatment in the 'powdering tub of infamy', a direct reference to the fumigant method. The fragment from Sonnet 153, 'seething bath, which yet men prove / Against strange maladies a sovereign cure' may be another reference to mercury vapour treatment. In fact,

* Some versions of the text have 'Nell' instead of Doll. See also Chapter 3.

there are so many Shakespearean references to venereal disease and its remedies that some have suggested the playwright may have been writing from personal experience.

Mercury might have been successful in treating the initial symptoms of syphilis and the lesions that appear in the tertiary phase of the disease, but it would have had no beneficial impact on secondary syphilis and the risk of poisoning was significant. The side effects of mercury treatment included a sore mouth and throat, sometimes involving ulceration, copious salivation (producing three pints of saliva was often considered a good sign), nausea and frequent bowel movements. In extreme cases it was the treatment that killed rather than the disease (see Chapter 3).

A milder but no more effective treatment came from a tree found growing in Hispaniola, the supposed origin of the disease: Guaiac wood. Guaiac wood was easily confused with ebony, hence 'the juice of cursed ebony' in *Hamlet*, which gives a very different interpretation to what killed Old Hamlet (see Chapter 8).

Most cures focused on the male sufferers and little comfort or care was offered to the women. Some even thought that women couldn't contract syphilis, but this was patently false as many women died of the disease. Thomas Nashe estimated that prostitutes were syphilitic by the time they were 20, and before they were 40 they were skeletally thin. Women that survived their time as prostitutes often went on to run their own establishment, as does Mistress Overdone in Shakespeare's *Measure for Measure*. Once a 'fresh whore', she has become a 'powdered bawd'.

One of Shakespeare's most memorable victims of syphilis is Pandarus. At the end of *Troilus and Cressida* it becomes clear that he has been infected with disease and doesn't have much time left. With only two months to live he appeals to his fellows in the sex trade (implying the audience):

> Good traders in the flesh, set this in your painted cloths.
> As many as be here of pander's hall,
> Your eyes, half out, weep out at Pandar's fall;
> Or if you cannot weep, yet give some groans,
> Though not for me, yet for your aching bones.
> Brethren and sisters of the hold-door trade,
> Some two months hence my will shall here be made:
> It should be now, but that my fear is this,
> Some galled goose of Winchester would hiss:
> Till then I'll sweat and seek about for eases,
> And at that time bequeath you my diseases.

He will spend his final weeks trying treatments to ease his pain, but he knows he will die. In his last line he bequeaths his fatal illness to the audience.

CHAPTER EIGHT

Most Delicious Poison

> If you poison us, do we not die?
>
> *The Merchant of Venice*, Act 3, Scene 1

Shylock, who speaks the words above, is right to be concerned about the lethality of poisons. In Shakespeare's time highly toxic compounds were readily available, while effective medical treatment and antidotes were in very short supply. Poison was a terrifyingly easy way to kill someone and because forensic knowledge was almost non-existent there was a good chance of getting away with it.

Autopsies were carried out to look for signs of poisoning, but unless a corrosive substance was used there would have been few characteristic signs that sixteenth-century medical expertise could have recognised. There was, however, one sign that would have been considered powerful evidence at the time: 'If they had swallow'd poison, 'twould appear / By external swelling' (*Antony and Cleopatra*). It was an accepted fact at the time that poisons caused the body to swell, but this has since been proved entirely false. There must have been several examples where a body, swollen by natural decaying processes, was mistaken for a poison victim and this may well have led to some wrongful convictions.

There was such a poor understanding of how toxic substances interacted with the body that it also led to the extraordinary situation where the same substance could be sold as a cosmetic, a medicine and a poison. Mercury, lead and arsenic-based compounds were swallowed for their supposed health benefits and smeared on faces as make-up, without any apparent alarm being raised by customers. Even when the dangers and ill-effects were well known there was no regulation and little concern.

Many women, and probably actors playing women's roles too, exposed themselves to damaging levels of heavy metals on a daily basis. The fashion of the day, set by Queen Elizabeth I, who was known to 'pile on the paint', was for very pale skin and red cheeks – ''Tis beauty truly blent, whose red and white / Nature's own sweet and cunning hand laid on' (*Twelfth Night*). White lead, in a form known as ceruse, was used to cover the face for the perfect pale complexion. It corroded the skin but this only encouraged users to apply an extra layer to cover the blemishes; hair also fell out and gums receded, loosening the teeth. Mercury, in the form of cinnabar, a brilliant red pigment, was used as rouge. It could lead to memory loss, paranoia and a grey film on the teeth. Both mercury and lead damaged nerves, causing headaches, depression and the shakes. Casual, everyday, low-level poisoning seems to have been taken in the average Elizabethan's stride. Rather than warning people off the use of these substances, fashions changed to accommodate the damage being done to the body – wigs and black teeth became all the rage.

Medicine was the other main route to unwittingly poison the general population. In an era when the aim of medical remedies was to restore the balance of the four humours (see Chapter 3), purgatives were in common use. White arsenic, or arsenic trioxide, with its powerful vomit-inducing properties, has therefore been used in medicine since as early as 2000 BC. Many other dangerous substances would have been considered effective treatments simply because of the body's natural reaction of vomiting them out after swallowing.

In *King Lear*, when Regan is killed by her sister Goneril with 'medicine', her symptoms are entirely what you would expect for an overdose of an Elizabethan remedy. Regan describes her condition: 'My sickness grows upon me.' She is 'Sick, O, sick!'

While people may have been accepting of mild everyday poisoning, deliberate poisoning was very different. Poison was, and still is, seen as a particularly underhand method of murder. Unlike suddenly lashing out in the heat of an argument, poisoning takes time and planning. There are lots of opportunities for the potential poisoner to pause, reflect and reconsider their actions before it is too late. Nor does the victim have an opportunity to defend himself or herself. It is perhaps then no surprise that those found guilty of poisoning are often considered the worst kind of murderer and are singled out for extra punishment. The sixteenth century was no different.

For example, in 1531, Richard Roose, a cook for the Bishop of Rochester, added poison to his master's evening meal. Two of the Bishop's guests died and several others were left with permanent ill-health. The Bishop survived but only because he wasn't feeling hungry that day. Roose claimed, after being tortured on the rack, that he had added laxatives to the meal as a joke, but no one else saw the funny side. His punishment was to be boiled to death in a cauldron.

* * *

Poison would seem to offer a lot of dramatic potential for the Elizabethan stage. There are certainly many references to poisons and poisonings dotted throughout Shakespeare's work. The word 'poison' appears over 130 times, 'venom' more than 40, and he scattered the names of poisonous plants and animals throughout his plays. In *Hamlet* alone lethal substances are poured into ears, smeared on swords and dissolved in drinks.

Using toxic substances to murder in a play has certain dramatic and practical advantages. There is no need for blood and gore to be used, which is always messy and a nightmare for the laundress to get out of the costumes. Actors also get to make the most of their death scene with choking, maybe some twitching or even convulsions. Such displays might not always have been scientifically accurate but it is convenient shorthand to let an audience know a character has been poisoned. Melodramatic endings like this are not always popular. One commentator wrote of the eighteenth-century actor David Garrick's drawn-out demise in *Hamlet*, 'we are not fond of characters writhing and flouncing on carpets'.

Modern audiences may love a good poisoning, as can be evidenced by the continuing popularity of Agatha Christie stories. But the fashion in Shakespeare's day was for exuberant displays of swordsmanship, and the playwright gave the audience what it wanted. The majority of Shakespearean deaths occur at the end of a sword or dagger. Relatively few characters die from being poisoned. What is surprising, given how knowledgeable Shakespeare was on a variety of subjects, is how poorly planned out his poisonings are. The Bard was brilliant at many things, but toxicology was evidently not one of them. Nevertheless, the poisons and poisonings included in his work are very interesting.

★ ★ ★

Shakespeare's knowledge of poison is patchy. If he describes symptoms, he rarely names the poison that caused them. On the other hand, if he names the poison, he rarely discusses how it will affect someone. For example, he names one poison 'ratsbane' to show the vindictive nature of some of his characters, without ever mentioning the symptoms. In *King Lear*, Edgar, who has disguised himself as Poor Tom, complains how others have treated him:

> Who gives anything to poor Tom? whom the foul fiend hath led through fire and through flame, through ford and

> whirlpool, o'er bog and quagmire; that hath laid knives under his pillow and halters in his pew, set ratsbane by his porridge …

Shakespeare's 'ratsbane' (rat poison) was probably arsenic, the poison made notorious over the centuries because of its frequent use to murder. It gets several mentions from the playwright, and all of them are to show the malicious intent of a character. In *Henry IV* Part I the Shepherd curses his adopted daughter, Joan la Pucelle (Joan of Arc), 'Now cursed be the time / Of thy nativity! I would the milk / Thy mother gave thee when thou suck'dst her breast, / Had been a little ratsbane for thy sake'. And in *Henry IV* Part II Falstaff quips, '[I] had as lief they would put ratsbane in my mouth as to offer it with security'.

Shakespeare's knowledge of poisons extends beyond arsenic. The highest concentration of poisonous substances is to be found in the witches' cauldron in *Macbeth*. The three weird sisters take it in turns to add ingredients to the pot. The first begins:

> Round about the cauldron go;
> In the poison'd entrails throw.
> Toad, that under cold stone
> Days and nights has thirty-one
> Swelter'd venom sleeping got,
> Boil thou first i' the charmed pot.

Poisoned entrails and venom are obviously not good for anyone's health, but these tantalising toxic details could be a subtle reference to the Borgias. Some members of the Borgias, a powerful Spanish-Italian family, were alleged to have been poisoning people around the turn of the sixteenth century using their own special preparation named '*La Cantarella*'. The recipe called for a pig (some accounts say a bear) and a lot of arsenic. The arsenic was first used to kill the pig/bear and then the poisoned entrails were sprinkled with more arsenic and left to rot down until a soupy mess was produced. Then the water was gently evaporated off to leave a pale powder. This powder was the *Cantarella*, which was reportedly added to various meals served to the Borgias' enemies. If this poison recipe seems elaborate, it has nothing on *Macbeth*'s witches' brew. The second sister continues:

> Fillet of a fenny snake,
> In the cauldron boil and bake;
> Eye of newt and toe of frog,
> Wool of bat and tongue of dog,
> Adder's fork and blind-worm's sting,
> Lizard's leg and owlet's wing,
> For a charm of powerful trouble,
> Like a hell-broth boil and bubble.

None of the second sister's contributions are particularly toxic. If the addition of snake parts was supposed to be poisonous, the 'Adder's fork' (its tongue) was the wrong bit. Shakespeare, however, does seem to have been under the impression that snakes delivered their venom using their forked tongue, as shown in *Edward III* – 'Let creeping serpents hid in hollow banks / Sting with their tongues'. But even if they weren't strictly poisonous, none of the ingredients sounds very pleasant and most of the animals included have negative associations. The third sister is clearly the medical and herb expert.

Scale of dragon, tooth of wolf,
Witches' mummy, maw and gulf
Of the ravin'd salt-sea shark,
Root of hemlock digg'd i' the dark,
Liver of blaspheming Jew,
Gall of goat, and slips of yew
Silver'd in the moon's eclipse,
Nose of Turk and Tartar's lips,
Finger of birth-strangled babe
Ditch-deliver'd by a drab,
Make the gruel thick and slab:
Add thereto a tiger's chaudron,
For the ingredients of our cauldron.

Two ingredients on the list, hemlock and yew, are well known for their toxicity. Hemlock's toxic notoriety comes from its use in ancient Greece, particularly in killing Socrates in 399 BC. Shakespeare names the plant three times in his works, but usually as a weed that easily seeds and takes over, rather than as the source of anything deadly. Yew, on the other hand, in the six times Shakespeare names it, is always associated with death. The trees often grow in English churchyards, and are therefore associated with the dead in the same way as cypress trees are in continental Europe.* But, unlike cypress, yew can also *cause* death as every part of the plant, except the flesh of the red berries, is toxic. The wood of the yew tree was also used to make bows and so Shakespeare is able to talk 'Of double-fatal yew' in *Richard II* to show the seriousness of the threat from an army.

* In *Twelfth Night*, Feste names both trees in his song about death.

One of the additions to the cauldron is not so easily recognisable as a poison. 'Tooth of wolf' may be just that, or it could refer to ergot, a black fungus resembling the shape of a fang that grows in cereal crops. Eating the fungus can cause blackening of the fingers, due to vasoconstriction cutting off the blood supply, as well as psychological effects.* In the past ergot was responsible for whole villages experiencing mass hallucinations when people ate bread made from contaminated flour. It was also used for centuries in herbal medicine to induce birth,† and today derivatives from the fungus are still used, safely, in conventional medicine to control post-partem bleeding.

The three sisters also express knowledge of specifically how and when ingredients should be collected, and it is true that levels of toxicity in a plant can change over the growing season. But the methods they describe, collecting at night, whether intended to minimise or maximise toxicity, would not be effective.

The witches' concoction is intended as a 'charm', but to what purpose is not said. Though the three sisters are clearly up to no good, they don't seem to be out to deliberately kill anyone. They may be encouraging Macbeth in his murderous acts, but they are not supplying him with poison to carry them out. The contents of their cauldron, however, would probably be an effective poison if anyone were rash enough to drink it.

---

* LSD was developed from compounds found within ergot.

† At considerable risk to the mother and child, it might be added, as the crude preparation cut off blood supply to the baby rather than specifically inducing birth.

Poisons can of course be beneficial, in the right circumstances and dose. As Paracelsus, the sixteenth-century physician and father of toxicology, said, 'it is the dose that makes the poison'. Or, as Shakespeare put it in *Henry IV* Part II, 'In poison there is physic'. Substances that interact with the body can modify a process in a way that corrects a fault. The same substance given in excess can change that process dramatically, or stop it all together, so that the body cannot function properly. For example, a drug that slows the heart rate can be useful in treating a heart that beats too fast. Too much of the same drug can slow the heart until it stops.

The potential dual nature of some substances was certainly understood by Shakespeare, even if the mechanism by which they acted was not. For example, Friar Laurence in *Romeo and Juliet* comments, 'Within the infant rind of this small flower / Poison hath residence and medicine power'. He is collecting materials that might be used as medicines but he is well aware of their potential dangers:

> O, mickle is the powerful grace that lies
> In herbs, plants, stones, and their true qualities:
> For nought so vile that on the earth doth live
> But to the earth some special good doth give,
> Nor aught so good but strain'd from that fair use
> Revolts from true birth, stumbling on abuse …

The Friar is able to put his knowledge to good use when he gives Juliet the potion that will make her appear dead (see Chapter 3). Such detailed medical knowledge is not unsurprising for a friar. Religious men were among the most highly educated groups of people in Shakespeare's day and were often looked to for medical help.* Friar Laurence did what he thought best for both Juliet and Romeo; he couldn't have foreseen the disasters that awaited them. Other religious figures in Shakespeare's plays are not so benevolent and use their knowledge of plants and potions for deadly effect – 'The king, I fear, is poison'd by a monk'. The king in question is John and his death is one of the few poisonings that Shakespeare describes in detail.

★ ★ ★

As King of England, John was in a position of great power, but in the play he is portrayed as a weak monarch who perhaps does not have the strongest claim to the throne. He therefore has reason to worry that someone might try to poison him and sensibly employs a taster. Unfortunately he chooses poorly; it is his taster who poisons him. It might be imagined that a monk would be a trustworthy individual, ideal for such a role, but the monk in question is 'a resolved villain'.

In an extraordinary detail it is said that the monk's 'bowels suddenly burst out'. This may be another way of saying 'spilling his guts' or confessing to the poisoning. It might also refer to a very physical symptom such as extreme vomiting or diarrhoea, perhaps from tasting his own poison. Another possible explanation is post-mortem swelling of the body. This swelling is most likely

---

* The witches in *Macbeth* could also be an unflattering portrayal of wise women, another common source of medical expertise at the time.

in the stomach area, where bacteria in the gut continue to feed on the individual's last meal and then the individual himself. The result of this feeding frenzy is rapid replication of bacteria and the production of gas. The abdomen inflates and if the gas builds up enough pressure, and there is no alternative outlet, bodies can rip apart suddenly and violently. What the phrase almost certainly *doesn't* mean is that poison literally caused his intestines to explode – no poison does that, not even in Shakespeare.

The monk never appears onstage to explain why he poisoned the monarch or how he did it. But King John is given considerable time to describe his symptoms in graphic detail, giving us clues as to what he might have swallowed. There is a 'burning quality', 'raging', 'strange fantasies' or hallucinations, and the King describes his 'bowels crumble up to dust', 'my burn'd bosom', 'my parched lips' and 'The tackle of my heart is crack'd and burn'd'.

Despite the abundance of information it is difficult to identify exactly what poison might cause such effects. White phosphorus poisoning gives a burning sensation and unquenchable thirst, but the element was unknown in Shakespeare's day. Blistering agents such as cantharides, sometimes known as 'Spanish fly' though the poison comes from a beetle, damage skin on contact. Strong acids or chemicals such as lye (see Chapter 2) can cause chemical burns – these are chemical reactions with skin and other tissues that can result in serious, even fatal, damage. Any of these would certainly explain the burning sensations, but not the hallucinations and the bowels crumbling to dust.

Perhaps a more likely candidate is aconitine. Symptoms of aconitine poisoning include a characteristic burning sensation in the mouth and throat, tingling and numbness in the skin, nausea, vomiting, chest pain and shallow breathing, convulsions and finally death because of respiratory failure or ventricular fibrillation. The poison comes from *Aconitum*, commonly known as monkshood, a species of plant that grows throughout Europe and is the most toxic native plant in the UK.* The toxic properties of *Aconitum* were certainly known to Shakespeare: the plant is named in *Henry IV* Part II and compared with gunpowder in terms of the damage it could do to the body. It still doesn't explain King John's hallucinations, but then the poisoner could have used a mixture of several poisons.

Another toxic plant native to the UK is *Atropa belladonna*, or deadly nightshade. Atropine, the toxic component found mostly in the berries of the plant, interacts with nerves. The result is an increased heart rate and drying up of secretions such as sweat, saliva and digestive juices. The interaction with nerves can also, in about half the population, cause visual, realistic hallucinations (as opposed to the psychedelic colours and patterns experienced with LSD). A combination of aconitine and atropine would easily explain all King John's symptoms.

Alternatively Shakespeare may have selected a range of symptoms purely for artistic effect. But he certainly didn't invent King John's poisoning itself. Holinshed's

* A single gram of plant material can kill, or a mere two milligrams of pure aconitine, if swallowed. The roots have been mistaken for wild radishes, with fatal results.

Chronicles, Shakespeare's source for the play, offered several possible explanations for the monarch's death, including poisoning by a monk. The real cause of death, however, is most likely to have been dysentery.

★ ★ ★

Monks and wise women were clearly knowledgeable about the production and use of poisonous substances. Such specialist knowledge was not available to everyone, thankfully. When Romeo needs to get hold of a poison in *Romeo and Juliet*, he has to consult an expert, and he chooses an apothecary.

Romeo is clear what he wants:

> let me have
> A dram of poison, such soon-speeding gear
> As will disperse itself through all the veins
> That the life-weary taker may fall dead
> And that the trunk may be discharged of breath
> As violently as hasty powder fired
> Doth hurry from the fatal cannon's womb.

But the apothecary is equally clear that selling poison is against the law, a tricky line to walk when most medicines at the time were also highly toxic. Nevertheless he is a poor man and Romeo's generous offer of 40 ducats sways him. The fact that he doesn't have to prepare anything and can instantly produce a bottle of poison meeting the criteria suggests, rather worryingly, that there was already a demand for such things and Romeo's request was not the first or even that unusual. And although the apothecary does not give a list of ingredients, the lethality of the liquid is spelled out in no uncertain terms: 'Put this in any liquid thing you will, / And drink it off; and, if you had the strength / Of twenty men, it would dispatch you straight.'

Whatever is inside the bottle, it is fast-acting and very potent, even when diluted. The apothecary's dire warnings are confirmed later when Romeo drinks it. The poison takes effect the instant he swallows it and he only has time to gasp out two lines and give Juliet a quick kiss before dying. Few poisons act so rapidly, and of those, fewer still were known about in the sixteenth century. The most likely candidate is cyanide, which could be extracted from a number of plant sources including peach or apricot stones and laurel leaves.[*]

If it was cyanide, Juliet's attempt to kill herself using traces of poison that were still on Romeo's lips could have worked. Cyanide compounds react with stomach acid to produce hydrogen cyanide, a gas, which can easily escape through the mouth and kill someone who kisses the lips of a cyanide victim, whether romantically or through attempts to give mouth-to-mouth resuscitation. A case in the USA illustrated this when a young couple, engaged to be married, sat next to each other on a sofa to discuss their wedding plans. The man took a piece of chewing gum from a packet and started to chew, but the gum contained a lethal amount of cyanide. The couple were found dead still sitting next to each other, the woman having died after kissing her beloved.[†]

---

[*] Nicotine would act with similar rapidity, but tobacco was relatively new to Europe and probably too expensive for a poor apothecary.

[†] The poisoner and the reason for the poisoning was never discovered.

Cyanide may be swift but it isn't pleasant. Inside the body, cyanide bonds to the active site of the enzyme cytochrome oxidase; this enzyme is crucial for converting glucose and oxygen into energy inside the cells of the body. If cyanide is present, no matter how much oxygen the body breathes in, the enzyme simply cannot process it. If a cell can't produce energy it rapidly dies. Cyanide kills because it causes massive cell death. Cells that require the most energy, such as nerves, are the most vulnerable and die first, causing headache, dizziness and convulsions, as well as vomiting and rapid pulse, before collapse and death. Any and all of these symptoms can occur within minutes. Romeo shows no interest in what the poison is; he simply sees it as a means to a tragic end. A more detailed knowledge might have dissuaded him from taking it.

Romeo may be uninterested in the poison's effects but other characters were not so happy in ignorance. Wise women, monks and apothecaries had legitimate reasons to investigate toxic substances in their professions. But when kings and queens start investigating poisons and their effects, their reasons are more likely to be malicious.

★ ★ ★

The Queen in Shakespeare's *Cymbeline* dedicates herself to researching all manner of chemical methods of killing so that she can murder those who stand in her way. Her intended victims are her husband the King and stepdaughter Imogen. She is the prototypical wicked stepmother familiar from so many fairy tales.

The Queen employs a doctor, Cornelius, to tutor her in how to synthesise her own perfumes and other preparations. What starts as lessons in basic chemistry techniques soon escalates into the study of poisons. Her excuse is that she simply wants to expand her knowledge, but her insistence that she only intends to kill small animals and not humans is hardly reassuring. Her reasoning is, 'To try the vigour of them and apply / Allayments to their act, and by them gather / Their several virtues and effects.' This is an unpleasant but scientifically reasonable way of determining the effects of different substances. To test an antidote for a poison you have to administer the poison as well to see if it works. And, as discussed earlier, what may be considered a poison in one context, in appropriate quantities, can be beneficial to health in another context.

The Queen uses a medical pretext to try to fool not only the doctor but also her intended victim, Imogen. She hands her stepdaughter the poisons in a box she claims is full of medicines. The doctor, however, has not been reassured by the Queen's talk of 'virtues' and beneficial effects. He may not be aware exactly what her plans are, but he is uneasy and wise enough to sabotage them. He tells the audience of his suspicions in an aside – 'I do suspect you, madam; / But you shall do no harm.' In place of the lethal poisons he substitutes drugs that 'Will stupefy and dull the sense awhile' rather than kill. When Imogen swallows the contents of the box, thinking they are mild medicines, she is knocked unconscious and the Queen's plans are thwarted.

Later, not realising the poisons had been swapped and Imogen lives, the Queen 'did confess she had / For you a mortal mineral; which, being took, / Should by the minute feed on life and lingering / By inches waste you'. Arsenic, lead or mercury-based compounds could all be considered a type of 'mortal mineral', though none of them would induce unconsciousness or a death-like appearance. The cautious doctor must have substituted something soporific in their place rather than simply diluting the mineral the Queen had requested (see Chapter 3 for some possibilities).

Cymbeline is rarely performed today and is perhaps one of Shakespeare's least known plays. The malevolent monarch has been eclipsed in the public consciousness by another poison-obsessed royal figure, Cleopatra.

★ ★ ★

The real-life Egyptian Queen had a reputation for her extensive toxicological knowledge – 'She hath pursued conclusions infinite of easy ways to die'. She fits many of the stereotypes of a poisoner: calculating, cunning and female. Poison may have a reputation as a woman's weapon, but this is without foundation.* Shakespeare has more male poisoners than female.

The playwright's knowledge of Cleopatra's history came from Plutarch's *Lives of the Noble Greeks and Romans*. The Greek biographer asserted that the Egyptian Queen had made a study of poisonous substances. She is said to

* Though it is true that a larger percentage of female murderers use poison than male, the number is still very small. There are far more male murderers than female, meaning that overall, male poisoners outnumber the women.

have built up a collection of poisons and experimented with them on prisoners condemned to death. Her experiments led her to conclude that poisons that acted quickly also caused extreme pain and convulsions. By contrast she believed that milder poisons were slower to act. She used the same methods to investigate bites from venomous creatures. According to Plutarch she watched these experiments personally. Her research may have been relatively scientific in its approach, but the methods were cruel and her conclusions were wrong.

Plutarch distinguished between venoms and poisons, although many others use the words interchangeably, particularly Shakespeare. In fact they are very different things. A poison is a toxic substance that is capable of causing the death of a living organism when introduced or absorbed. A venom is a special type of poisonous substance, one secreted by an animal specifically as an act of aggression, rather than in defence.

In defence an animal only needs to distract the predator long enough to escape, usually by causing pain. In predation, on the other hand, the venom has to kill or disable the victim rapidly before it can get away. Venoms have a lot of work to do and quickly, so they are often, but not always, more complex mixtures than those used for defence. Venoms can include any and all of the following categories of substances: salts, peptides, proteins (such as enzymes), lipids and amines. All of these compounds will have a potential role in disabling the victim. For example, potassium salts can induce pain by causing nerves to fire. Enzymes operate as molecular machines that can rapidly carry out chemical reactions and, in the case of those in venoms, may destroy blood vessels and tissue, cause blood clots or prevent blood from clotting, as well as a host of other damaging effects on the body. Members of the amine class of compounds can act as neurotransmitters also affecting nerves.

How lethal a venom is depends on several factors – the toxicity of the components, how much can be delivered by the animal in a bite and where the victim is bitten. Toxicity is often measured by a number known as the LD_{50}, the quantity required to kill half of a number of test animals. How much venom is delivered in a bite or repeated biting depends both on the animal and on the situation. The site of injection is also important. Some venoms are more toxic when injected into a vein or the abdominal cavity. Others are more lethal when injected under the skin or into muscle.

According to Plutarch, Cleopatra's investigations into venoms led her to believe that the bite of an asp was the best option for suicide as it brought on a gradual lethargy, 'in which the face was covered with a gentle sweat, and the senses sunk easily into stupefication'. Given the complexity of different venoms and how they act, it is no surprise she got things a bit wrong. If she had wanted a gentle, painless death it is unlikely that she got her way using an asp.

Roman and Greek writers referred to asps in a way that suggests the word was a common name for several different species. The most common species of snake in Egypt is the Egyptian cobra, *Naja haje*, and is the snake usually associated with ancient Egyptian culture. It measures 1–1.5m (3–5ft) in length and can deliver a lethal bite. Its venom contains mostly neurotoxins and cytotoxins (substances toxic to cells). The neurotoxins prevent signals from being sent between nerves, resulting in several symptoms. There will be a lot of pain at the site of the bite as well as in the abdomen. The effect on nerves that control muscles means there will be a slow flaccid paralysis, which is perhaps what Cleopatra observed in her experiments. When the paralysing effects progress to the heart and lungs, death results from complete respiratory collapse, but there is likely to be dizziness and convulsions before that happens. The effects of the cytotoxins will be seen as severe swelling, blistering, bruising and necrosis (cell death).

All of this presents considerable problems when it comes to Cleopatra's famous suicide. For a start, the snake was supposed to have been brought to Cleopatra in a basket of figs. But the basket would have been far too small to contain a snake big enough to kill.* The death itself is also difficult to explain based on the accounts left to us.

Shakespeare has Cleopatra describe the bite as 'sweet balm' and 'soft as air', but this is far from what it is likely to have felt like. Egyptian cobra bites are painful enough as it is, but her choice of where to be bitten possibly made things worse. Different parts of the body have different numbers of nerve endings detecting pain, so it is theoretically possible to map out the most painful places using a good knowledge of nerve distribution. One researcher went a step further and tested the theory with practical experiments. In 2014, Michael Smith

* One account reported in Plutarch says the snake was carried into the room in a water vessel and that the Queen had to torment it with a needle before it would bite.

allowed himself to be stung five times by honey bees each morning. He then rated the pain on a scale of 1–10. Over a period of weeks he was stung in 25 different parts of his body. The most painful place to be stung, he concluded, was the nostril, which he scored 9.0.*

Shakespeare has the Queen apply the snake to her breast, though probably more for reasons of artistry than for historical accuracy, followed by another bite to the arm.† According to Smith's pain scale, the nipple pain level rated 6.7 out of a possible 10. The top of the forearm scored 5.0 and the wrist 4.7. Cleopatra would have been better off persuading the snake to bite the skin of her skull or the tip of her middle toe, both of which score a mere 2.3 on Smith's scale.

After being bitten, Plutarch reports that the Queen died very quickly, too quickly for those who ran to help her. This rapid death is copied by Shakespeare in his play; there is scarcely time for a dozen lines between the bite and her death. Charmian, one of Cleopatra's attendants, also applies an asp to her own arm but doesn't even make it through seven lines of dialogue; she just has time to make adjustments to Cleopatra's diadem (perhaps the convulsions knocked it askew), before she collapses.

A guard enters six lines after the death, too late to be of any help. Even if someone had been there on the spot, instantly sucking the venom from the wound, it would

---

* Smith was awarded an Ig Nobel prize for his work. The prize has been awarded to scientists since 1991 to 'honor achievements that first make people laugh, and then make them think'.

† Plutarch also claims the marks of a snake bite were found on Cleopatra's arm.

not have saved her. Venoms spread quickly through the body. It is unlikely anyone would be able to suck hard enough to remove the toxic substances. Likewise, you wouldn't be able to cut into the flesh with a knife quickly and deeply enough to stop the venom from spreading.* And while the spread of venom through the body is indeed very rapid, death from the venom of the Egyptian cobra is not. The effects of the toxins within the venom take time to develop and kill.

In the play, as more people enter the scene they find not only Cleopatra and Antony, but also two attendants, 'All dead.' With no obvious sign of how the women died, investigations are made. Poisoning is suspected but dismissed instantly because there is no sign of swelling on any of the bodies. But closer inspection of Cleopatra reveals 'Here, on her breast, / There is a vent of blood and something blown: / The like is on her arm', perhaps the mark of the bite itself together with the bruising and blistering effects of the venom. Further evidence of a snake is discovered: 'This is an aspic's trail: and these fig-leaves / Have slime upon them, such as the aspic leaves / Upon the caves of Nile'.†

The search for evidence is only cursory, and the bodies of the attendants, Charmian and Iras, don't even get a glance. Had they been examined, the bite marks on Charmian's arm would have been discovered, but Iras's body would have presented a considerable puzzle. There

---

* The best advice, should you find someone who has been bitten by a snake, is to apply pressure to the wound and call an ambulance.

† Shakespeare clearly knew very little about snakes – they do not leave a slimy trail.

would be no marks to find. She dies instantly after kissing Cleopatra on the lips. At this point the Queen hadn't yet been bitten by the snake, so it can't be a mistaken belief that toxins in the body from a snake bite are somehow transferable by a kiss. There must be some other explanation for Iras's death; perhaps she died of grief or distress (see Chapter 10).

The inconsistencies between the accounts of Cleopatra's death and the reality of death by cobra venom have created a lot of doubt in people's minds. Plutarch also acknowledged these doubts and wrote about several alternative theories. He included one account that said no snake was found in the room, but that reptile tracks were apparently found on the sea sands opposite the windows of her apartment. The reason for assuming she had been bitten by a snake may be because it was a symbol commonly used by the kings of Egypt and would have been depicted on Cleopatra's diadem. Another theory Plutarch put forward was that she had in fact used poison that she kept concealed in a hollow bodkin worn in her hair.

Any confusion over how exactly Cleopatra died has nothing to do with Shakespeare or his poor knowledge of venoms. In *Antony and Cleopatra* he was dramatising historical events and any errors are from mistakes or misunderstandings in his sources. He didn't need to invent anything; the iconic death was already full of drama before Shakespeare put it onstage. In other plays he used his imagination more freely.

★ ★ ★

Hamlet contains more poisonings than any other play (at least three different poisons contributing to five deaths), but what poisons were used on which characters has left many people scratching their heads. The most famous poisoning in the play, that of Old Hamlet, is described in great detail and the poison used is even named. But despite the apparent wealth of information it is still not clear what killed him.

According to the account related to Hamlet by his father's ghost, Old Hamlet was deliberately poisoned by his brother Claudius. Whether Hamlet is hallucinating or he really sees the departed soul of his father is a debate for another book. If we assume what Hamlet hears is a true account of events, there is a huge amount of detail about not only what Claudius did, but how he hoped to get away with it.

Sleeping within my orchard,
My custom always of the afternoon,
Upon my secure hour thy uncle stole,
With juice of cursed hebona in a vial,
And in the porches of my ears did pour
The leperous distilment; whose effect
Holds such an enmity with blood of man
That swift as quicksilver it courses through
The natural gates and alleys of the body,
And with a sudden vigour it doth posset
And curd, like eager droppings into milk,
The thin and wholesome blood. So did it mine;
And a most instant tetter bark'd about,
Most lazar-like, with vile and loathsome crust
All my smooth body.

> Thus was I, sleeping, by a brother's hand
> Of life, of crown, of queen, at once dispatch'd …

The poison is named as hebona, but no such poison seems to exist. Spelling was something of a movable feast until it was standardised in the eighteenth century. Before then words were constructed out of letters based on pronunciation and personal taste. So perhaps in *Hamlet* the playwright is just using an unusual spelling for something else. Possibilities include hemlock, hellebore, henbane or ebony.

First, the least convincing possibility: hemlock. Even if Shakespeare took extreme liberties with the spelling, it is a stretch to get to hebona from hemlock. The symptoms of hemlock poisoning, either dizziness and convulsions or a creeping paralysis, depending on which of the two poisonous hemlock plants you are talking about, are also very unlike those described by Old Hamlet.

The second alternative is hellebore. In Pliny's *Natural History*, written in the first century AD and a work certainly known to Shakespeare, there is a mention of these plants growing among vines and the wine produced from them causing abortions. Pliny also knew that the plant could kill horses and pigs if they ate it. But in humans it is most likely to cause a severe upset stomach rather than death.*

Henbane, the third possibility, is closer both in spelling and in possible symptoms. *Hyoscyamus niger*, commonly known as henbane, is one of the Solanaceae family of plants that contains a staggering number of toxic members, including jimsonweed and the deadly nightshade that

* Even so, please don't eat hellebore or feed it to other people.

we met earlier in the chapter. These plants, and several others in the same family, contain atropine, which switches on the body's 'fight or flight' response. Pupils dilate, and secretions, like sweat and saliva, dry up. One other symptom that is sometimes observed is a rash, particularly on the upper torso, though this is not exactly the 'vile and loathsome crust' Old Hamlet complains of.

Another connection in support of henbane is a reference in Pliny's *Natural History*. The book contains a recipe for curing earache using the juice of henbane, opium and rose-oil among other ingredients. The mixture was warmed up and introduced into the ear using a syringe. It would also explain Shakespeare's strange choice of the ear as the site for applying the poison.

Poisons are usually ingested or injected in some way. A few can be inhaled and some can be absorbed through the skin. However, the ear is a particularly poor choice for application. The inside of the ear is protected with wax, making absorption difficult. There are also relatively few blood vessels in the ear that can absorb the poison into the body proper. However, if Pliny is anything to go by, pouring things into the ear was a more common way of administering drugs than might be imagined, though with no obvious benefits and probably considerable discomfort. However, inserting something into the ear, like a syringe or tube, to deliver the substance could have perforated the eardrum and allowed a poison or medicine easier access to the rest of the body.

The final possibility, ebony, or *hebenus* or *hebeno* as it was sometimes referred to, may on the face of it seem less convincing. The spelling is very similar, but ebony is not particularly poisonous and was not considered to be so in the sixteenth and seventeenth centuries. But there is another type of wood that was often confused with ebony at the time, from the tree *Guaiacum officinale L.* This tree grows in the West Indies and on the north coast of South America and was named by the natives of Hispaniola, who called it *guaiacum* or *guaiacan*. Another name given to the tree was 'pockwood' because extracts of the wood were often used to treat the pox or syphilis (see Chapter 7).*

This offers an alternative interpretation of Old Hamlet's revelations to his son. He is perhaps accusing his wife of infecting him with syphilis contracted from her affair with his brother. There are other allusions to syphilitic infection in this passage. Old Hamlet says the poison caused his skin to become 'bark'd', 'crusted' and 'lazar-like', terms often used in the past to describe the secondary stages of syphilis when an all-over rash of pustules appeared. The poison, or the hebona juice, is also described as a 'leprous distilment', and in Shakespeare's day leprosy was often used as an all-encompassing term for diseases that affected the skin. Further, the poison is said to run through his veins 'swift as quicksilver' – an alternative name for mercury and perhaps a reference to mercury treatments for syphilis.

Syphilis is far from a definitive interpretation of this passage in the play, but there are enough references to the disease for Shakespeare's audiences to have picked up on them. Later on, when Hamlet confronts his mother, he makes further suggestive remarks: 'Such an act [...] takes off the rose / From the fair forehead of an innocent love, / And sets a blister there' may be an allusion to the

* Ineffectively, it might be added.

first signs of syphilitic infection, a sore or chancre. Hamlet also uses the word 'burn' in a way to suggest painful urination, a symptom of venereal infection. Other phrases he uses, such as 'panders will' and 'Stew'd in corruption', would be associated in the audience's minds with brothels, the common source of syphilitic infection.

Of course, Shakespeare may not have meant any of this and could have invented a poison to suit his theatrical needs. But the effects seem so specifically described that it is hard to believe he was not referring to something real. He might also have simply borrowed the name from somewhere or someone else. Christopher Marlowe used the name 'hebon' for a poison he mentioned in *The Jew of Malta*, a play written long before *Hamlet*, but Marlowe included no information about symptoms or origins of the toxic substance.

If Shakespeare didn't borrow the idea from Marlowe, maybe it was from the original story of Hamlet. The tale comes from a Scandinavian legend first recorded around 1200 AD by the Danish historian Saxo Grammaticus in his *Gesta Danorum*. But this version has the old king killed by a snake bite. In Shakespeare's version this is the story put about by Claudius to explain Old Hamlet's sudden death, as the ghost explains: ''Tis given out that, sleeping in my orchard, / A serpent stung me.' In *Hamlet* the snake story seems to be generally believed, but is it a credible explanation for Old Hamlet's untimely death?

There certainly is a venomous snake native to Denmark, the European asp (*Vipera aspis*). The venom from this snake can produce symptoms including rapidly spreading acute pain, which would certainly fit the description of 'swift as quicksilver it courses through / The natural gates and alleys of the body'. This is followed by swelling from oedema (excess fluid accumulating in the cavities and tissues of the body). Blood vessels and tissue are degraded, causing severe necrosis. The venom has both coagulant and anticoagulant effects causing significant changes to the blood, perhaps something like 'it doth posset / And curd, like eager droppings into milk, / The thin and wholesome blood'. It was certainly a believable explanation for Old Hamlet's symptoms and death. It is not surprising Claudius got away with it for so long.

Hamlet's method of exposing Claudius's guilt, by getting a visiting acting troupe to act out a play featuring an identical method of murder, is unconventional. It is more like Poirot's revelations at the end of an Agatha Christie novel than modern methods of acquiring evidence from forensic examinations and trials in courts of law.* And though, as we would hope, the guilty are punished, Hamlet's inactions after exposing his uncle's guilt characterise the play. He is pressured to take revenge, but completely fails to do so, resulting in a lot of innocent deaths along the way.

It all comes to a head in the final scene of the play when Hamlet and Laertes are set to fight. Hamlet believes he has the advantage because he has been practising his fencing skills, but he doesn't know the odds are stacked against him. Claudius has conspired with Laertes, and the tip of his sword has been laced with poison: 'I'll touch my point / With this contagion, that, if I gall him slightly, / It may be death.' And, to

* Indeed, Christie borrowed the title of *Hamlet*'s play within a play, *The Mousetrap*, for her most successful stage play.

make sure Hamlet will be killed, even if he wins the match, Claudius promises to add another poison to the winner's drink.

These are both rather more conventional methods of poisoning. Shakespeare doesn't even bother to name the substances used. Whatever they are, they are fast-acting. Cyanide might be a likely candidate for the cup. What is added to the sword is more difficult to guess. Whatever it is it must be lethal in very small amounts and swift in its action. Aconitine, mentioned earlier in association with King John, is one possibility as extracts of the plant were used as an arrow poison in ancient times.

Another possibility is curare, a nerve toxin used to tip arrows in central and southern America. It causes paralysis and kills because the lungs can no longer expand to take in oxygen. It is only toxic when introduced into the bloodstream and can be safely eaten, meaning hunters could eat their kill without fearing they would also be poisoned. This exotic arrow poison was first heard about in England in 1596 after Sir Walter Raleigh returned from his explorations in the Americas, though it is possible the poison he wrote about wasn't curare.

Claudius's and Laertes's plan works, up to a point, but they clearly haven't thought through all the possibilities. Such fast-acting and potent poisons leave others vulnerable to their effects. Their actions result in four deaths rather than the hoped-for one. Gertrude, innocent of the plot, drinks a toast to her son from the poisoned cup and dies soon after. Laertes receives a wound from the poisoned sword and becomes another victim. Hamlet is wounded as expected and the poison starts to take effect. Before he succumbs he stabs Claudius with the poisoned sword. What poisons were used is not important in this scene. The point is the acceleration of action, the death after death that builds to the climactic finish, contrasted with the sudden stillness of Horatio who, of all the main characters in the play, is the only one left alive to tell their tragic tale.

CHAPTER NINE

To Be, or Not to Be

Out, out, brief candle!

Macbeth, Act 5, Scene 5

The first line of Hamlet's famous soliloquy, 'To be, or not to be,' is perhaps the most quoted of any in Shakespeare's entire works. *Hamlet*, more than any other play, explores the theme of suicide in detail, from the Prince of Denmark's soul-searching soliloquies to Ophelia's drowning, and the different responses to her death.

Hamlet's contemplation of life, and his deliberations over whether death could bring the blessed relief and escape he craves, is a heart-wrenching exploration of the darkest moments in a person's life. He shows all the typical signs that might lead up to the act itself: depression and anxiety, suicidal thoughts and even planning stages. His mother's deep concern for her son's welfare prompts her to enlist friends to talk to him and support him. It is what anyone would do for a friend or family member.

The intervention is successful and Hamlet does not commit suicide.

The circumstances that drive Shakespeare's characters to commit the ultimate act are extreme and few people today would expect to find themselves in a similar situation. But the emotions – grief, fear, desperation – are real enough. The playwright's incredible insights into the bleakest human experiences are unnervingly accurate. His fictional characters have provided decades of psychologists with case studies. The portrayal of suicidal characters in the plays and poems is not only deep in understanding but also surprisingly modern in attitude.

Shakespeare went a considerable way to show a more compassionate view towards those that contemplate and bring about their own death than might be expected for the time in which he was writing. The historical stigma of suicide was considerable. Natural death was one thing, something to be expected and accepted as part of life. Deliberately causing one's own death, however, was considered most unnatural and brought shame on the individual and their family.

In the past, to contemplate such a drastic act a person had not only given up hope of life, but a profound belief in the afterlife meant they had also given up all hope of the hereafter – 'the dread of something after death, / The undiscovered country, from whose bourn / No traveller returns' (*Hamlet*). Suicide was a grave sin, something Hamlet is evidently concerned about when he contemplates his own suicide, 'that the Everlasting had not fix'd / His canon 'gainst self-slaughter!' The remains of suicides were buried in unconsecrated ground, at crossroads, or in fields, and with a stake driven through their body. And, as if that weren't enough punishment, suicide and attempted suicide were criminal offences – all the victim's money and possessions would be confiscated by the state.

Contrary to what might be expected in Elizabethan society, Shakespeare's suicidal characters are not shunned and ignored; they are mourned and pitied. The double suicide of Romeo and Juliet, brought about by a series of simple misunderstandings; Eros, 'Thrice-nobler than myself!' who kills himself rather than his friend Antony; or Othello, who realises too late that his wife was innocent and kills himself over the guilt of her murder, are all usually seen as tragic wastes of life. They are figures of sympathy rather than shame.*

In the majority of cases, had the character in question waited a little longer, their circumstances would have changed and misunderstandings would have been corrected. Hope and a happier life is just around the corner, but the character can't see it. Signs that can lead to suicide are also missed and opportunities to help are lost, due either to simple ignorance or to society's negative attitude.

Much has changed since the sixteenth century. Public opinion is overwhelmingly sympathetic and there is much more help available. Those contemplating suicide can be more open about their feelings without being ostracised. Support organisations that offer sympathy and

* Despite Shakespeare's more tolerant view of suicide he seems to have had little impact on society's perceptions. Suicide and attempted suicide were only decriminalised in the UK in 1961.

help can be easily reached, such as counselling services, Samaritans or suicide hotlines. Today, talking to friends and loved ones is most likely to elicit compassion and care rather than condemnation. Such opportunities were not available to Shakespeare's characters, even though in some cases suicidal tendencies are noticed and attempts made to help and prevent them. Studies have shown that one method of preventing suicide is restricting access to the means of carrying it out, which is a tactic employed in a few of Shakespeare's plays.

⋆ ⋆ ⋆

In the periods when he set his plays, from the sixteenth century stretching back to ancient Roman times, methods of suicide were inefficient and agonising.

> To die – to sleep –
> No more; and by a sleep to say we end
> The heartache, and the thousand natural shocks
> That flesh is heir to – 'tis a consummation
> Devoutly to be wish'd. To die – to sleep.
> To sleep – perchance to dream: ay, there's the rub!

Hamlet's description of death as like falling asleep may have been what he wished for, but the reality was usually very different.

More than a dozen characters commit suicide in Shakespeare's play and poems, and more than half do so by stabbing themselves, three of these occurring in just one play (*Julius Caesar*). In the majority of these cases the playwright was following the historical record, and these lessons from history show that to fall on your sword is not necessarily a swift and noble death. It is a dramatic but painful and violent method that does not always result in the rapid end many may be hoping for. Shakespeare was horribly aware of the miserable reality of self-inflicted stab wounds.

It takes a lot of determination to commit such a violent act on your own body in the full knowledge of the pain that it is going to cause. The three deaths in *Julius Caesar* are a case in point. Two of the three suicides, Cassius and Brutus, have to enlist help from devoted friends to carry out the act, by requesting their friend stab them or hold the sword while they fall on it. In *Antony and Cleopatra*, Antony, believing Cleopatra is dead, tries to enlist the help of his friend Eros to kill him. But Eros kills himself rather than Antony. His sacrifice to save Antony's life actually spurs Antony on. 'Thrice-nobler than myself! / Thou teachest me, O valiant Eros, what / I should, and thou couldst not.' Copycat suicide is not a new phenomenon.

The situation with Eros and Antony also shows how self-stabbing can easily be botched. Antony falls on his sword but fails to give himself the swift dramatic death he craves: 'How! not dead? not dead?' His sword likely failed to pierce a vital organ or major blood vessel. Though he begs those around him to 'give me / Sufficing strokes for death', they refuse. He collapses in agony, slowly bleeding to death. In another horrifying twist, he then learns that Cleopatra is still alive and his dramatic gesture has been for nothing. Unable to stand, he has to be carried to her to say his farewell. It is not until the next scene and 140 lines later that Antony eventually succumbs to his wound.

The chances of surviving a stab wound, whether self-inflicted or otherwise, depends largely on what is damaged by the blade. Stab wounds to the abdomen, for example, can bleed relatively slowly if they miss the spleen and major blood vessels, and death can be delayed for several hours. Wounds to the intestines may be survived initially, thanks to modern surgical techniques, only for peritonitis to set in that can lead to life-threatening infections throughout the whole body.

Stabbing the chest, the target area for all of Shakespeare's self-stabbings, might be expected to be more rapidly fatal, but it is not straightforward. The vital organs within the thoracic cavity are protected by the rib cage. A sharp blade can easily penetrate the intercostal muscles and cartilage between the ribs but the injury that results can vary.

Bleeding into the chest cavity (haemothorax) can be considerable if a major blood vessel or the heart has been punctured. Surprisingly there might not be much external bleeding, as when the knife is withdrawn the tissues can overlap and close up like a valve. The victim can bleed to death internally with little sign of the damage from the outside.

In other cases a hole can be made connecting the interior and exterior of the chest, causing a 'sucking wound'. When the victim breathes, instead of drawing air through the nose and mouth, air enters into the chest directly through the wound.

It is a 'sucking wound' that is the likely cause of death for the title figure in Shakespeare's long narrative poem *The Rape of Lucrece*. Lucrece stabs herself in the chest over shame at her rape. The injustice of her situation is explored at length: 'O, let it not be hild / Poor women's faults, that they are so fulfill'd / With men's abuses: those proud lords, to blame, / Make weak-made women tenants to their shame.' Her rapist is racked with guilt and punished by banishment. He loses the right to enter the city of Rome, but Lucrece loses her life. The disparate treatment of the two is noted: 'To slay herself, that should have slain her foe.'

Lucrece's plan, 'against my heart / Will fix a sharp knife to affright mine eye; / Who, if it wink, shall thereon fall and die', does not work. Her life doesn't end in a wink. The witnesses to her suicide stand and watch apparently stunned into inaction. Little help is offered to the dying woman, but then there was little that could have been done to save her. The horror of the rape, and the wound Lucrece inflicts on herself, shocks not only those around her that witness the stabbing, but the reader as well.

The detailed description of her wound emphasises the brutality of her treatment at the hands of her rapist. Blood 'bubbling from her breast' suggests that air is escaping through the wound and that she pierced a lung rather than her heart as she had intended. The line 'this fearful flood' shows bleeding was extensive and the likely cause of death in this case.

This poem also demonstrates Shakespeare had made some detailed observations of the behaviour of blood. 'Some of her blood still pure and red remain'd, / And some look'd black' is a direct acknowledgement of how blood can change its appearance. Though the black colour is attributed to tainting from her rapist's sin, the line 'Congealed face of that black blood' shows the poet also understood there could be a physical explanation for the change. As the blood comes into contact with the air, it coagulates and darkens. The words 'watery rigol' describe how congealed blood can separate into clots and serum.

The fact that both red and blackened, fresh and congealed, blood is present around Lucrece's wound would indicate she remained alive for some time after the stabbing. She would have been in considerable pain and struggling to breathe throughout.

As with many of Shakespeare's suicides, Lucrece's death is lamented: 'Do wounds help wounds, or grief help grievous deeds?' – there is no attempt to cover up the manner of her death and in fact it prompts her family to seek justice.

Even if the heart isn't always specifically mentioned, it is the target in Shakespeare's self-stabbing cases – 'Come, Cassius' sword, and find Tintinius' heart' (*Julius Caesar*). It would seem to be the most obvious way to achieve a rapid death. What is not so obvious is that the organ is remarkably resilient to stab wounds.

First of all, the heart is well protected by the rib cage, and the breast bone in particular. It takes considerably more force to penetrate bone with a blade than muscle or cartilage. The blade must also be long enough and angled correctly to reach the heart. But even then things may not go as swiftly as might be expected.

There are several examples from history of stabbings directly to the heart that have not proved immediately fatal. For example, in one case in the twentieth century, a homicide victim was stabbed *through* the heart but still ran more than a quarter of a mile to chase down his attacker before collapsing. The damage observed in post-mortem examinations is not an accurate predictor of how long a person can survive stabbings to the chest. Comparisons of eye-witness accounts of stabbings with pathology results would back up Shakespeare's observations of both rapid and prolonged deaths from stab wounds.

Some parts of the heart are more vulnerable than others. The left ventricle of the heart has a thick wall of muscle that, when it contracts, can partly seal a wound. Life can continue while blood leaks into the pericardium (the sac surrounding the heart). At a certain point the pericardium will become full of blood. The pressure of the liquid means the heart can no longer expand and death occurs (a situation described clinically as 'cardiac tamponade'). Wounds of the right ventricle, however, are often more rapidly fatal as the thinner wall is not so effective in stopping blood from escaping. In contrast, damage to the major blood vessels that enter the heart is rarely survivable. Today, emergency treatment can stem blood flow, transfusions can compensate for blood loss and modern surgical techniques can attempt to repair the damage if the individual is reached in time. But little or nothing could be done in the past and it was simply a matter of waiting for the inevitable.

Many of the self-stabbings in Shakespeare's works are based on historical fact or well-known stories. The Roman characters Brutus, Tintinius, Cassius, Cato, Mark Antony and Eros were all real people, and they all died by self-inflicted wounds. The characters Romeo and Juliet, Pyramus and Thisbe (depicted in *A Midsummer Night's Dream*) and Lucrece, were popular from poems and prose works long before Shakespeare placed them on the stage. Depicting self-stabbings onstage was not only faithful to the original stories but created a visually arresting and dramatic moment. It is also relatively easy to portray. Swords could be safely slipped under arms to avoid injury and bladders of blood could be pierced to allow blood to flow (see Chapter 2).

This is perhaps one reason why Shakespeare chose self-stabbing for Othello's death even though this was not in the original story. In Giovanni Battista Giraldi's tale, *A Moorish Captain*, the Othello character did not commit suicide. He escaped from the prison where he was being punished for his part in his wife's murder, only to be killed by her relatives to avenge her death. Another reason is that by changing who killed Othello, Shakespeare has the character acknowledge his guilt and portrays him in a more sympathetic light. It also increases Iago's guilt, as Othello's suicide was ultimately the result of his actions. It is a much more complex depiction of human behaviour and motivations than the original story.

Other methods of suicide Shakespeare explored in his plays were not so easy to stage and did not always follow the accepted real-life events he dramatised. It is interesting to speculate why he chose some of them. One suicide in particular presents problems in terms of staging, certainly as far as the health and welfare of the actor involved is concerned. The stage directions for Arthur in *King John* appear to require the actor to fall from a height in full view of the audience.

★ ★ ★

King John is perhaps Shakespeare's least performed play. Its relative obscurity means the plot and characters are not as familiar to audiences as others. One of the main themes of the play is sovereignty: whether it is Arthur or John who should wear the crown. Outside the walls of Angiers, English troops are massed in support of John and a French army is also there in support of Arthur. Angiers remains resolutely neutral in the whole affair. There is an inconclusive battle and all sides are forced back into negotiations. A lot of discussion, a strategic marriage and a treaty between the two sides manages to avoid another war, but only temporarily. John is accepted as King of England but relations between the two nations break down when John defies an instruction from the Pope. France (supporters of the Pope) and England (supporters of their king) find themselves in opposition again. War ensues, during which Arthur is captured by the English.

The young Arthur is entrusted to the care of Hubert de Burgh, in reality a powerful administrator John had sent to France to assist in the wars, but in the play more of an evil henchman carrying out John's orders. One of those orders is to kill Arthur. With Arthur out of the way, John believes his claim to the English throne will be stronger than ever.

At Rouen Castle, Hubert means to kill Arthur by putting out his eyes with hot irons. But the boy's pleas melt his captor's heart and Arthur's life is saved. Hubert returns to John and confesses that he has failed the King, and Arthur still lives. But, while Hubert is away making his confession, the young boy takes the opportunity to disguise himself and try to escape.

He stands on the castle walls contemplating the drop in front of him. Arthur does not intend to die, but understands it is a likely outcome of his actions: 'I am afraid; and yet I'll venture it. / If I get down, and do not break my limbs, / I'll find a thousand shifts to get away'. He reasons that if he dies in the attempt he still wouldn't be in a worse position than he is now, i.e. likely to die by John's orders – 'As good to die and go, as die and stay.'

Arthur's wish, 'Good ground, be pitiful and hurt me not!' does not come true. The fall is fatal but he does not die straight away. His injuries are severe, but he survives long enough to speak two more short lines.

Studies by NASA have shown that falling from a height of six metres (20ft) onto a hard surface and landing on your feet is likely to cause major trauma. From this height the body will be travelling around 25 mph when it hits the ground. But thanks to modern medicine such a fall is probably survivable. Arthur was living, and dying, in the thirteenth century, without any practical medical help. At a height of 7 to 12 metres (23–40ft) survival is questionable. Anything above 12 metres and onto rocks is almost certain death.*

The height of a fall does not always correlate with the damage sustained. Some people can be killed by falling from a standing height, while others may jump from a great height and escape without injury, usually because of some cushioning effect.† The type and severity of the

* The terminal velocity of a human in the pike position (bent over at the waist) is roughly 200 mph.

† Staging falls from a height requires experts and detailed planning to be carried out safely. Stunt performers train and rehearse to carry out their work but there are still risks.

injuries largely depends not only on the height of the fall but also on which part of the body hits the ground first.

Controlling a fall is very difficult. Even if Arthur simply allowed his body to drop, falling from a height and aiming to land on his feet, it wouldn't have been simple. Depending on the height, and other factors, the body can tumble as it falls, meaning any part of the body might be pointing downwards when it eventually hits the ground. Bodies can also bounce and ricochet on their way down.

Landing from a fall on the top of the head is rare, even from a considerable height. The thickness of skulls varies from place to place as well as between individuals but there is likely to be a massive fracture. Landing on the feet is more likely, and damage can occur not only to the bones in the legs, but the legs can also be forced upwards through the pelvis. The force of impact can then be transmitted up the spine causing damage at any, or all, points along the route, even all the way up into the skull.

If the fall is onto the side, any combination of injuries can occur. Ribs, arms and shoulder bones can be broken and internal organs, such as the liver, lungs, heart and spleen, can be lacerated or ripped from their moorings. When the chest is suddenly decelerated, the aorta connecting the heart to the rest of the circulatory system can be ruptured, leading to massive haemorrhage that will kill quickly.

Injuries can vary enormously from one case to the next. Some bodies can be severely fragmented but in some cases the skin can be almost intact, obscuring the severe disruption that has occurred to the internal organs. Post-mortem examinations have shown that it is almost impossible to determine the height of a fall from the nature and severity of the injuries.

How Shakespeare imagined that Arthur's drop would be depicted onstage is difficult to say. Stage directions are kept to a bare minimum: '[*He leaps down*]'. There are no details as to how this might be achieved safely. Surely the actor in question has to fall on something cushioned to avoid the same fate as Arthur. He could jump out of sight of the audience but the actor then needs to be seen to deliver his final two lines and his body must remain in place to be discovered by a group of English lords.

However it might be achieved, it needn't be a realistic depiction. As has been discussed before in Chapter 2, realism was not necessarily the goal of Elizabethan theatre. An emotional response from the audience can be provoked in many ways: by Arthur's defiant speech, for example, and the lords' laments when they discover the body. It is the desire to escape and Arthur's bravery in attempting it, in full knowledge of the possible consequences, which are the important aspects of the death, not the manner of it. An audience doesn't need to see a broken body to appreciate the tragedy of a young man's death.

When Arthur's dead body is discovered by the English lords they speculate over the circumstances of his death – did he jump or was he pushed? Hubert interrupts the scene. His initial jubilation that he has done the right thing by saving the boy swiftly gives way to despair when he finds the lords mourning over Arthur's body. Hubert is briefly accused of having pushed the boy off the wall, but it is impossible to prove one way or the other. Hubert's evident distress at Arthur's death, however, counts in his favour. Even today evidence to determine accident, suicide or murder at the scene of a fatal fall is difficult to obtain and often relies on circumstantial evidence such as blood-alcohol levels, suicide notes and testimony from friends and family over the victim's state of mind.

Determining how the real-life Arthur died is just as difficult, owing to a lack of clear forensic evidence and reliance on rumour. Holinshed's *Chronicles*, Shakespeare's source for the play, states he was to be killed by having his eyes put out, but was saved by Hubert de Burgh. From that point onwards, Shakespeare and the historical record diverge. The *Chronicles* note several theories of Arthur's fate, including one that he attempted to escape by climbing down the walls of Rouen Castle and jumped into the Seine, where he drowned. Another theory says the boy was consumed with grief and pined away, dying of a natural sickness. Yet another version of events, said to come from the commander of Rouen Castle, claims that agents were sent by King John to castrate Arthur and he died of shock from the bungled surgery.

Most of these accounts were written by those who didn't like Arthur and wanted to discredit him. It may be more likely that Arthur died of disease when in prison. Shakespeare seems sympathetic to the young man and gives him more choice over his end than the real-life Arthur may have had.

A large part of the tragedy of Arthur's death in the play is that he is so young. It is an uncomfortably accurate scenario that echoes real life, as most suicides are committed by adolescent boys and young men. To compound the horror, Shakespeare emphasises the futility of his death. Had Arthur waited a little longer he would have heard the news from Hubert that King John wanted him to live.

Shakespeare may have chosen an impractical way to stage Arthur's death, but the emotional impact is huge. However, it isn't always necessary to show the final moments of a character to elicit sympathy from an audience. Portia's horrific death in *Julius Caesar* is a case in point.

* * *

Portia spends little time onstage, but when she does she is clearly preoccupied and unwell. From her very first appearance in the play, her health is described as poor – as her husband Brutus says, 'It is not for your health thus to commit / Your weak condition to the raw cold morning.' Evidence of her distressed state of mind comes when she reveals to Brutus that she has been self-harming: 'Giving myself a voluntary wound / Here, in the thigh'.

A few scenes later she complains of feeling faint with worry over her husband. She sends a servant to find out news of him. The last we see of her is outside her house not wanting to worry her husband but desperate to hear news of his meeting with Caesar: 'Run, Lucius, and commend me to my lord; / Say I am merry: come to me again, / And bring me word what he doth say to thee.' The play moves on and Portia, worried and in poor health, is temporarily forgotten.

In the second half of the play, the action has moved away from Rome to the site of battle between forces loyal to Brutus and those loyal to Mark Antony. While

Brutus is sitting in his tent planning his campaign against Mark Antony, he announces to Cassius, 'Portia is dead.'

Such a bald statement takes Cassius, and the audience, by surprise. Brutus has to repeat himself, 'She is dead.' Cassius's first thought is that she must have been sick, but Brutus explains:

> Impatient of my absence,
> And grief that young Octavius with Mark Antony
> Have made themselves so strong: – for with her death
> That tidings came; – with this she fell distract,
> And, her attendants absent, swallow'd fire.

The dialogue then moves on to other topics. Just when it seems Portia is to be forgotten again, news of the death is repeated a third time. Some have suggested that it is an error when the play was printed, but when it is acted out onstage it is very effective. Cassius, struggling to come to terms with what he has been told, repeats 'Portia, art thou gone?' showing how difficult it can be to accept such sudden and devastating news. Brutus in reply pleads 'No more, I pray you.' A similar pattern is used in *Antony and Cleopatra* when Antony receives news that his wife Fulvia has died after a long sickness. The news has to be repeated several times before it sinks in.

The manner of Portia's death is particularly shocking. Swallowing fire was not an invention of Shakespeare's for dramatic purposes. Plutarch, the Greek biographer, wrote:

> As for Porcia, the wife of Brutus, Nicolaüs the philosopher, as well as Valerius Maximus, relates that she now desired to die, but was opposed by all her friends, who kept strict

> watch upon her; whereupon she snatched up live coals from the fire, swallowed them, kept her mouth fast closed, and thus made away with herself.

However, Plutarch was not convinced the account was true. Modern historians also doubt the veracity of the method. Swallowing hot coals, if it is even physically possible, is an extreme act. Death would be likely to be caused by the swelling of the damaged tissues in the throat closing the airway and suffocating the poor woman. The pain would have been excruciating. Instead it has been suggested that Portia burned coal or charcoal in a closed room and died of carbon monoxide poisoning.

Whatever the method, it shows how unwell Portia must have been. It seems that those around her knew, or at least suspected, her intentions and did their best to prevent it. In *Hamlet*, Ophelia also shows potentially suicidal behaviour, but it appears to go unrecognised. Her brother Laertes seems more concerned with exacting revenge on Hamlet, the person he blames for his sister's madness, rather than helping Ophelia.

⋆ ⋆ ⋆

Ophelia's death is perhaps the most famous of all in Shakespeare's canon. Her drowning, even though it takes place offstage for obvious reasons of practicality, has been imagined and immortalised in poems, songs, and paintings ever since. The event itself may not be shown to Shakespeare's audiences but it is described in detail by Gertrude:

> There is a willow grows aslant a brook,
> That shows his hoar leaves in the glassy stream.
> There with fantastic garlands did she come
> Of crowflowers, nettles, daisies, and long purples,
> That liberal shepherds give a grosser name,
> But our cold maids do dead men's fingers call them.
> There on the pendant boughs her coronet weeds
> Clamb'ring to hang, an envious sliver broke,
> When down her weedy trophies and herself
> Fell in the weeping brook. Her clothes spread wide
> And, mermaid-like, awhile they bore her up;
> Which time she chaunted snatches of old tunes,
> As one incapable of her own distress,
> Or like a creature native and indued
> Unto that element; but long it could not be
> Till that her garments, heavy with their drink,
> Pull'd the poor wretch from her melodious lay
> To muddy death.

The build-up to this moment is considerable. Hamlet's erratic behaviour towards Ophelia and his cruel rejection comes to a crisis when he murders her father. Ophelia is shown distracted, barely aware of her surroundings and talking repeatedly about her father's death, acting out a mock funeral to an imaginary audience. Her behaviour is a 'document in madness!'

Gertrude's account of Ophelia's death suggests it is her distraction and unawareness of her surroundings that is the cause rather than a deliberate attempt to kill herself. Her interest in flowers and singing snatches of old tunes is a continuation of her previous behaviour. It seems she barely even notices that she is in danger and makes no effort to save herself.

Today, Ophelia's death would be mourned as a tragic accident and might prompt investigations into whether something could have been done to prevent it. In the sixteenth and early seventeenth centuries the discussion would have been slightly different and focused on her intentions.

Drowning was the most common method of self-destruction for women in Shakespeare's era. It was important to establish whether the act had been intentional or not in order to give an appropriate burial for the body. The decision would be made by the coroner, referred to in *Hamlet* as the 'crowner'. If the death was by the person's own volition, the body would not be buried in consecrated ground and all their property would be seized by the Crown. Suicide as a result of madness made the death free of sin and conventional burial and inheritance was allowed, but this verdict was rarely given in the sixteenth century. In the play, the subject is discussed by two gravediggers:

> First clown: Is she to be buried in Christian burial when she wilfully seeks her own salvation?
> Second clown: I tell thee she is; therefore make her grave straight.
> The crowner hath sate on her, and finds it Christian burial.
> First clown: How can that be, unless she drown'd herself in her own defence?
> Second clown: Why, 'tis found so.

Gertrude, in her report of the death, emphasises both Ophelia's behaviour and the fact that a branch happened to break under her. However, when it comes to the funeral it is clear the priest is not convinced over the verdict, saying 'Her death was doubtful', and he won't perform the full ceremony. 'No more be done. / We should profane the service of the dead / To sing a requiem and such rest to her / As to peace-parted souls.' He makes it clear that he thinks more has already been done for her than she deserves: 'Shards, flints, and pebbles should be thrown on her.' Ophelia's burial in consecrated ground but with a minimum ceremony was a compromise only rarely offered in this era.

Accidental death does not seem to have been considered, despite Gertrude's assertions, but this would also have allowed a Christian burial. The difficulty then, as it is today, is in differentiating accident from suicide. In drowning cases there are no characteristic post-mortem signs that can be used to distinguish between the two. Circumstantial evidence, such as leaving notes or removing clothes and glasses before entering the water, is used to establish suicidal intention. But not every suicide behaves in this way.

Although there is debate about whether Ophelia fell by accident or willingly, there is no doubt that she drowned. Drowning is in effect a form of asphyxia. Water displaces the air that would normally flow into the throat and lungs. Suicidal drownings are less likely to fight against the sudden influx of water – Ophelia seems to have been very passive in her final moments. Her clothes buoyed her up for a while until they became soaked and helped drag her body under the water. She, like some people, may have been able to hold her breath for around a minute. Oxygen reserves within the body could delay brain death for a few more minutes. After that, even if her body was recovered from the water, she would be unlikely to survive.

Asphyxia is the usual cause of death in drowning cases but there are other important factors involved. Sudden immersion in cold water can cause intense stimulation of nerve endings just below the skin, inducing changes in normal heart rhythm and triggering cardiac arrest. Cold water entering the pharynx and larynx can stimulate nerve endings in the mucous membranes with the same result for the heart. The fact that Ophelia was picking flowers immediately before she drowned suggests spring or summer, so it was probably not cold enough for these effects.

Another possible mechanism of death when the head is submersed is that a gulp of water enters the trachea, causing a reflex cardiac arrest before there is time for lack of oxygen to take effect. This explains the deaths of individuals who have been pulled from the water very quickly after immersion. There is no indication of how long Ophelia was submerged before her body was recovered.

Not all drowning victims have water in their lungs. So-called 'dry-lung' drownings occur, it is thought, because of 'laryngeal spasms' causing closure of the airway. However, death in these cases is still due to lack of oxygen. But in most cases it is water entering the lungs that is the problem, and it isn't simply the displacement of air that causes changes in the body.

In the lungs, diffusion causes fresh water to be drawn into the bloodstream and dilutes the blood.* The extra

* Salt-water drownings cause the reverse process, where water is drawn out of the body to dilute the salt.

water can increase the blood volume by 50 per cent and in its diluted state it cannot carry out its normal functions effectively. The water also disrupts the blood itself by destroying red blood cells. Sometimes, in cases of so-called near-drowning, when the victim is pulled from the water and recovers, there is a later threat to life from water in the lungs (pulmonary oedema) and infection, particularly if the drowning occurred in brackish water.

In the play, Laertes' comment 'Too much of water hast thou, poor Ophelia,' is a very good way of describing many of the dangers of drowning. It also suggests some examination of the body, or maybe even an attempt at revival, that found evidence of water in the lungs. If attempts were made to save Ophelia, they were not successful – 'Alas, then she is drown'd?'

Lifeless bodies are denser than water and sink. The head is the densest part and will sink first. Ophelia would have remained under the water, most likely facing upwards, until someone could pull her body from the water. The most famous image of Ophelia's death, by the Pre-Raphaelite John Everett Millais, has inspired countless reproductions and imitations. These romanticised images show Ophelia placidly floating in a stream but still alive. The reality of her 'muddy death' would not make an appealing image.

Hamlet also shows the devastating effects of suicide on those left behind. Gertrude is in mourning, Laertes is in shock and Claudius is angry. Hamlet's sudden appearance at the funeral sparks a row between him and Laertes and they come to blows. There is blame and accusation on all sides. Everyone struggles to come to terms with the death.

Although Shakespeare's play is an adaptation of a much earlier story from the *Gesta Danorum*, he made considerable changes and additions to the plot. There is an Ophelia character in the original tale, but she does not go mad and does not kill herself. Shakespeare also made use of an earlier play by Thomas Kyd, known as *Ur-Hamlet*, but the play is lost and it is not known if it was Kyd who first introduced the idea of Ophelia's death by drowning.

Another possible source of inspiration comes from real-life events close to Shakespeare's Stratford home. In 1580 a young woman named Katherine Hamlett drowned in the river Avon. It happened near Tiddington at a junction in the river overhung with willow trees. The rumour was she had committed suicide but her family insisted she had fallen when she went to draw water from the river. The coroner must have agreed with the family as she received a Christian burial.

It has been speculated that Shakespeare could have worked as a town clerk in Stratford, where he might have heard of the case. He might also have heard about it from his father, who was an alderman in Stratford and would have been involved in investigating such cases as part of his civic duties. Wherever the story of Ophelia's tragic life and death came from, it is a haunting exploration of grief and suicide.

CHAPTER TEN

Excessive Grief the Enemy to the Living

But died thy sister of her love, my boy?

Twelfth Night, Act 2, Scene 4

Duke Orsino's question to Cesario about the fate of his sister is not one of incredulity at the cause of death, but a genuine concern that such intense emotion could potentially be fatal. Orsino was not alone in being worried about the dangerous effects of strong emotion. Several Shakespearean characters talk of the possibility – 'excessive grief the enemy to the living' (*All's Well That Ends Well*), 'Since that my beauty cannot please his eye, I'll weep what's left away, and weeping die' (*Comedy of Errors*).

According to the playwright they were right to be worried. Shakespeare's plays are littered with the bodies of characters that have died of grief, remorse, love or some other strong emotion. More than one character appears to die of a guilty conscience. For example, in *Richard II* the 'Abbot of Westminster, / With clog of conscience and sour melancholy / Hath yielded up his body to the grave'. Mostly, the reports of these deaths are taken at face value, details aren't asked for, and the validity of the statement is not questioned – it seems to be an accepted fact that such things can happen.

Shakespeare was writing at a time when physical well-being and emotion were wrapped up together in the humoural theory of health (see Chapter 3). According to the medical theories of the age, it might be expected that emotion could disrupt the normal equilibrium of the body to an extent that could kill. At a time when the true causes of disease and death were not well understood it could be that the Bard was in fact talking about some symptom of a physical disease that was misinterpreted, or he may have been exaggerating for dramatic effect. But it could also be that he rightly understood that strong emotions can kill.

Fortunately, Shakespeare's plays contain some clues as to what might be going on. Some characters have their symptoms described, offering an opportunity for modern interpretations of cause of death. And, though most of the characters that are killed in this manner are fictional, some were real figures from history. The historical record can be consulted to see if he was merely repeating the accepted explanation of the time or offering his own interpretation of symptoms and circumstances described by chroniclers.

★ ★ ★

The fact that emotion can have a physical effect on the body is no surprise. Fainting through fear or over-excitement is well known, from fans swooning at the sight of their idol, to passing out at the fear of needles or the sight of blood. The modern medical term for this is vasovagal syncope; the heart rate slows, blood vessels in the legs dilate and blood drains to the lower extremities. The drop in blood pressure and reduced flow of blood to the brain causes the pale complexion: 'he'll swoon! Why look you pale?' (*Love's Labour's Lost*) – and there is light-headedness and fainting – 'I should rejoice now at this happy news; / And now my sight fails, and my brain is giddy' (*Henry IV* Part II). In these cases there is normally a full recovery within minutes without the need for medical intervention.

There are more than two dozen examples of swooning or fainting in Shakespeare's plays. He has monarchs, ordinary folk, and men as well as women collapse from emotion or the sight of some horrible bloody scene, for example Lady Macbeth's collapse after the discovery of Duncan's murdered body. There are also a range of emotional triggers. In *The Winter's Tale*, Hermione swoons from grief at news of the death of her son – 'Her heart is but o'ercharged; she will recover'. In *King Lear*, blind Gloucester faints from fear when he believes he has fallen from a cliff edge. In *Pericles*, Thaisa faints with joy when reunited with her husband and daughter.

Thaisa's unconscious state is mistaken by Pericles for death. He isn't being melodramatic; mistaking a faint for death is common in Shakespeare's plays. A similar event in *Much Ado About Nothing* sees Hero faint, not from joy, but in response to an angry accusation of infidelity and threats from her father. Again, those gathered around her unconscious body believe she is dead. Fortunately for both Thaisa and Hero, they make a full recovery. Other characters are not so lucky.

Hero's situation is similar to a real-life case in Germany in 1856. A girl collapsed and died after being publicly berated by a teacher.* This was the first recorded case of 'long QT syndrome', a condition that causes irregular electrical activity in the heart. Mental or physical stress can create an irregular heartbeat resulting in sudden collapse with no other signs or symptoms. Without medical intervention after a collapse, death is likely. Long QT syndrome can be inherited but it can also be caused by certain medications.

Another cause of sudden unexplained cardiac deaths in young people is catecholaminergic polymorphic ventricular tachycardia (CPVT). It is a genetic disorder, first described in 1960, that affects proteins controlling the concentration of calcium in the heart – an element essential for allowing heart cells to contract and create a heartbeat. Exercise or emotional stress can lead to blackouts or even cardiac arrest. Both long QT syndrome and CPVT predominantly affect young people.

Either of these two conditions could explain the sudden death of Mamillius, the young son of Leontes in *The Winter's Tale*. Mamillius is still a child when he hears of the imprisonment of his mother and the news kills

* When the girl's parents were informed they told how another of their children, a son, had died after a fright, evidence that it might be an inherited condition.

him: 'The prince your son, with mere conceit and fear / Of the queen's speed, is gone.'

★ ★ ★

However, Shakespeare does not confine his emotional deaths to the younger generation. In *King Lear*, Gloucester, an older man who has already suffered considerable physical and emotional stress – he was tortured by having his eyes gouged out – is exiled and is later tricked into believing he is about to walk off a cliff edge. Gloucester is in an understandably fragile state when he is eventually reunited with his estranged son. The emotional stress of the reunion proves too much for him: 'But his flaw'd heart / (Alack, too weak the conflict to support!) / 'Twixt two extremes of passion, joy and grief, / Burst smilingly.' Edgar's report of the death gives some hint at a possible physical cause, a flawed heart.

Older people may be less likely to die from long QT syndrome or CPVT, but the effect of strong emotion and physical torture would have an understandably deleterious effect on their health. In the case of Gloucester perhaps a pre-existing heart condition could turn fatal after his harsh recent experiences. The reasons behind other adult deaths in Shakespeare's play are less well described, relying purely on emotional factors.

In *Romeo and Juliet* Lady Montague's death is reported by her husband: 'Alas, my liege, my wife is dead to-night; / Grief of my son's exile hath stopp'd her breath', but no further details are given. Her death is overshadowed by the tragedy discovered in the Capulet tomb. Similarly, in *Cymbeline*, Leonatus's death from grief is reported as a statement of fact: 'this gentleman in question, / Two other sons, who in the wars o' the time / Died with their swords in hand; for which their father, / Then old and fond of issue, took such sorrow / That he quit being'. In *Othello*, Brabantio dies because 'pure grief / Shore his old thread in twain'. Grief seems to be sufficient explanation for these deaths.

Another possible death from emotional causes is that of Enobarbus in *Antony and Cleopatra*. Although he dies onstage there is little indication in the dialogue, and no stage direction, to indicate precisely how:

> O sovereign mistress of true melancholy,
> The poisonous damp of night disponge upon me,
> That life, a very rebel to my will,
> May hang no longer on me: throw my heart
> Against the flint and hardness of my fault:
> Which, being dried with grief, will break to powder,
> And finish all foul thoughts. O Antony,
> Nobler than my revolt is infamous,
> Forgive me in thine own particular;
> But let the world rank me in register
> A master-leaver and a fugitive:
> O Antony! O Antony!
>
> [*Dies*]

Enobarbus is clearly very emotional and his death is usually attributed to remorse for betraying Antony, but there are no physical symptoms described. He does not commit suicide; the guards who find his body see no sign of injury and at first assume he is sleeping. It seems Enobarbus is simply able to wish his life at an end.

Shakespeare hadn't run out of ideas; he was being faithful to the true history of Gnaeus Domitius Ahenobarbus, or at least the version of events written in Plutarch's *Lives*, which Shakespeare used as the source for much of the play *Antony and Cleopatra*. Ahenobarbus was a politician, general and friend of Mark Antony, but he defected to Augustus's side at the Battle of Actium. He didn't fight though. He died, according to Plutarch, a few days after he reached Augustus from 'the shame of his disloyalty and treachery being exposed'.

Such a sudden death in an adult, without obvious signs or symptoms, might be explained by apical ballooning syndrome, a condition causing dangerous changes in the heart. In most cases it occurs after an incident of extreme mental or physical stress. Another name for the condition is takotsubo cardiomyopathy, after the Japanese word for an octopus pot. The shape of the octopus pot, with a narrow neck and a round bottom, is similar to the shape of the deformed left ventricle of the heart as it balloons. The left ventricle of the heart pumps oxygenated blood round the body and when it balloons it doesn't contract efficiently. The mechanism of how the heart can be physically altered by dramatic changes in emotion is still poorly understood.

Because the condition can be triggered by a recent sudden bereavement, it is more popularly known as 'broken heart syndrome'. Whatever you call it, the condition affects only a tiny percentage of the population. There are usually severe chest pains, breathlessness and collapse. Apart from grief, other triggers can also include domestic abuse, financial worries or debt, physical assault or illness, though occasionally no source of stress is identified. Most cases occur in women over 50, though it has been seen in much younger women and in men. The death of Lady Montague, though perhaps not yet in her fifties, would be a classic candidate for the condition. However, since its discovery it has been rarely shown to be fatal. The heart usually returns to normal within days or weeks without any treatment. Things might not have had such a good outcome before general health care could support the patient while they recover.

The deaths of Gloucester, Lady Montague and Enobarbus all occur rapidly with no extended period of change in health preceding their deaths, or certainly none that is mentioned. The suffering of other characters, however, is more prolonged. The behaviour of Doctor Pace in *Henry VIII*, Lady Constance in *King John*, and most famously Lady Macbeth in *Macbeth* changes dramatically owing to stress. In the plays their deaths are directly linked to their changed emotional state, or as young Lucius puts it in *Titus Andronicus*, 'Extremity of griefs would make men mad'.

* * *

In *Henry VIII*, Doctor Pace is a former secretary to the King. He never makes an appearance onstage, but his history is discussed between Cardinal Wolsey and Cardinal Campeius: 'you envied him, / And fearing he would rise, he was so virtuous, / Kept him a foreign man still; which so grieved him, / That he ran mad and died.' It is only a brief mention of what seems to have been a prolonged change in mental health brought on by stress that led to death.

In real life this character was Richard Pace, a diplomat who was sent on several important assignments in Europe before being recalled to England in 1526. He was simultaneously Dean of St Paul's, Salisbury and Exeter. In February 1536, it is reported that Pace was to be supported in his role as Dean by Richard Sampson because his 'mental imbecility for many years past has interfered with the due government of the cathedral'. He died a few months later at the age of 54. This may have been a case of early-onset dementia but Shakespeare's account of his death is certainly not inconsistent with the historical record.

Considerably more poetic licence was taken in *King John* where Lady Constance and her son Arthur are portrayed. Constance, Duchess of Brittany, really was the mother of Arthur, the claimant to the English throne in opposition to King John. He was only 15 when he led a campaign against his uncle but was captured and held prisoner at Rouen Castle where he is believed to have died (see Chapter 9). In Shakespeare's account, Lady Constance's agonised laments and despair over the separation from her son and his mistreatment are among the most memorable parts of the play.* Shakespeare probably wrote Constance's harrowing lines some time in 1596, the year Hamnet, his 11-year-old son, died:

> Grief fills the room up of my absent child,
> Lies in his bed, walks up and down with me,
> Puts on his pretty looks, repeats his words,
> Remembers me of all his gracious parts,

---

* Jane Austen, on a visit to London, was very put out when a performance of *King John* she had hoped to see was replaced with *Hamlet*. Sarah Siddons, the leading tragic actress of the day, was intended to have played Lady Constance. Instead Austen reluctantly went to see *Macbeth* performed the following Monday.

Stuffs out his vacant garments with his form;
Then, have I reason to be fond of grief?
Fare you well: had you such a loss as I,
I could give better comfort than you do.
I will not keep this form upon my head,
When there is such disorder in my wit.
O Lord! my boy, my Arthur, my fair son!
My life, my joy, my food, my all the world!
My widow-comfort, and my sorrows' cure!

Other characters in the play think Constance's grief has changed into something else: 'Lady, you utter madness, and not sorrow.'

To lose your sense of who you are was taken as a sign of madness. For example, in *The Taming of the Shrew*, Christopher Sly is tricked into thinking he is a rich nobleman. Not understanding why everyone around him calls him 'honour' and 'lordship', he protests, 'What, would you make me mad? Am not I Christopher Sly, old Sly's son of Burton Heath; by birth a pedlar, by education a cardmaker, by transmutation a bear-herd, and now by present profession a tinker?' In *King John*, Constance uses a similar tactic, knowledge of who she is, as proof she is not mad:

I am not mad: this hair I tear is mine;
My name is Constance; I was Geoffrey's wife;
Young Arthur is my son, and he is lost:
I am not mad: I would to heaven I were!
For then, 'tis like I should forget myself …

She may not be mad in any delusional sense but she is clearly distressed and suicidal: 'My reasonable part produces reason / How I may be deliver'd of these woes, / And teaches me to kill or hang myself'. Two scenes later a messenger reports that 'The Lady Constance in a frenzy died'.

If Constance had carried out her threats of suicide it might be expected her death would be reported as such. Instead the word 'frenzy' is used. The modern definition of frenzy – a period of uncontrolled excitement or wild behaviour – is different from the sixteenth-century understanding. The word is derived from 'phrenesis', described by Philip Barrough in 1593 as a 'continual madness' joined with acute fever – a physical as well as mental disorder. Although it might have several causes and manifestations (Barrough classified three types of phrenesis according to the mental functions that might be affected), it was a specific diagnosis and considered an incurable and deadly condition.*

Shakespeare may have been making a sophisticated medical diagnosis, but he was slipshod with his historical accuracy. The real-life Lady Constance died on 5 September 1201, almost a year before Arthur was captured and imprisoned. The cause of her death has variously been given as leprosy or complications during childbirth, but not 'frenzy'. In Holinshed's *Chronicles* Queen Elinor, King John's mother, is also said to have died of grief and 'anguish of mind'. In fact Elinor (Queen Eleanor of Aquitaine) died three years after Constance in her early eighties. She had supported John in his fight against Arthur for the crown, and after Arthur's capture (see

* Phrenesis fell out of favour as a medical term in the nineteenth century, to be replaced by delirium, confusion or clouding.

Chapter 9) returned to England and spent her remaining years as a nun. In the play, the deaths of Elinor and Constance are reported in the same sentence. Shakespeare deliberately or accidentally conflated the two.

He may also have felt they both needed a more dramatic ending and perhaps took inspiration from another real-life figure, such as Cardinal Beaufort. It is said the Cardinal died in a state of delirium, offering England's treasury to Death in return for living a little longer.* In *Henry VI* Part II, written before *King John*, Shakespeare depicts the death of Cardinal Beaufort more graphically:

> Cardinal Beaufort is at point of death;
> For suddenly a grievous sickness took him,
> That makes him gasp and stare and catch the air,
> Blaspheming God and cursing men on earth.
> Sometimes he talks as if Duke Humphrey's ghost
> Were by his side; sometime he calls the king,
> And whispers to his pillow, as to him,
> The secrets of his overcharged soul;
> And I am sent to tell his majesty
> That even now he cries aloud for him.

The explanation for the Cardinal's disturbing cause of death is given by Warwick: 'So bad a death argues a monstrous life.' It was thought that sin, or remorse for a sinful life, could have life-threatening psychological and physical effects. The line 'See, how the pangs of death do make him grin!' suggests the Cardinal's face shows a 'risus sardonicus' (a prolonged muscle spasm that looks like a sardonic grin). Considerable physical exertion is needed

* Other reports say he died quietly and with dignity.

to produce such an effect and it is usually seen during convulsions induced by tetanus or strychnine poisoning.

The possibility of grief, stress or guilt leading to prolonged anguish, even madness, and death perhaps stuck in Shakespeare's mind. He used a very similar idea, explored in much more detail, in a later play, *Macbeth*.

★ ★ ★

Lady Macbeth, following her involvement in the murder of Duncan, is somewhat preoccupied. Her guilt dramatically affects her mental health and behaviour. To outward appearances she is a calculating and cold-blooded murderer, but the facade crumbles to reveal a tormented soul.

Lady Macbeth encourages her husband to kill the Scottish King and so claim the crown for himself. She plans the murder with meticulous detail to ensure the blame will be laid elsewhere. Even immediately after the death she plays the part of incredulous bystander perfectly. It is Macbeth who is initially most affected by the death, not his wife. He can't sleep and begins to hallucinate. Lady Macbeth has to try to cover for her husband's obvious distraction. Only later, as the bodies pile up, do the physical effects on Lady Macbeth, brought on by her increasing psychological stress, start to show themselves – 'she is troubled by thick-coming fancies / That keep her from her rest.'

The preoccupation with sleep seen in *Macbeth* is characteristic of Shakespeare. Modern sleep researchers have praised his insights into sleep and speculated that the playwright wrote from bitter personal experience. His plays contain references to many sleep conditions, from insomnia – a disorder particularly affecting royalty in his plays: 'Uneasy lies the head that wears a crown' (*Henry IV* Part II) – to sleep apnoea, where breathing rapidly stops and starts – see Falstaff 'snorting like a horse' and 'Hark, how hard he fetches breath' in *Henry IV* Part I.

Lady Macbeth is perhaps the most famous literary example of the most well-known sleep disorder, sleepwalking (somnambulism), as well as sleeptalking (somniloquy). Stress and anxiety are acknowledged triggers for periods of disturbed sleep such as insomnia but sleepwalking is a rarity in adults. Those that suffer from it normally have a history of sleepwalking from their childhood onwards. How Lady Macbeth's emotional state affects her sleep, and how it ultimately leads to her death, is worth exploring more closely.

Shakespeare gives a detailed portrayal of her disrupted sleep. A doctor is called in by Lady Macbeth's gentlewoman to observe her as she sleepwalks. Even though she is asleep she manages to successfully light and carry a candle. The two observers watch her compulsively wash her hands and complain aloud about bloodstains on her hands: 'Out, damned spot!' and, 'Here's the smell of the blood still: all the perfumes of Arabia will not sweeten this little hand'. Although sleepwalkers are known to be able to carry out complex tasks like cooking or driving a car, more usually their behaviour is much less dramatic and meaningful than Shakespeare's example would suggest.*

Sleeptalking can vary from incoherent mumbles to long speeches. Most of Lady Macbeth's somniloquy is

---

* However, bed partners often report more elaborate behaviour than is observed in controlled environments such as sleep clinics.

fragmented repetitive phrases about blood and the stains on her hands. But there are moments when she appears to be thinking and reasoning. The phrase 'who would have thought the old man to have had so much blood in him' is more than an expression of horror and shows Lady Macbeth analysing Duncan's murder in some depth. In the four humours theory of health of the time, the human body was thought to slowly dry out as it aged. Older people, like Duncan, would not have been expected to bleed so profusely according to the medical theories of the day.

Another statement from the sleeping Queen, 'What need we fear who knows it, when none can call our power to account?' seems to be almost a confession. But it is less likely that sleeptalkers will reveal some deeply repressed secret in their night-time ramblings. Shakespeare's doctor is probably off the mark when he says, 'infected minds / To their deaf pillows will discharge their secrets'.

A few scenes later the doctor is discussing Lady Macbeth's restlessness with her husband. Macbeth asks the doctor to give her something:

> Cure her of that.
> Canst thou not minister to a mind diseased,
> Pluck from the memory a rooted sorrow,
> Raze out the written troubles of the brain
> And with some sweet oblivious antidote
> Cleanse the stuff'd bosom of that perilous stuff
> Which weighs upon the heart?

It almost sounds as though he is recommending psychotherapy to manage Lady Macbeth's sleep disorder.

If so, he is centuries ahead of his time. The modern approach to alleviating sleep disruption is to treat the underlying cause, be it a medical or a psychological issue. In the case of Lady Macbeth it is most certainly psychological. But her doctor cannot help and she dies. Could it be that her extreme emotional state had deprived her of sleep to the point that it killed her?

The idea that lack of sleep could kill can be found long before Shakespeare hinted at it. In Plutarch's *Lives*, a source for several Shakespeare plays, though not *Macbeth*, there is the story of Paulus Aemilius. He was said to have been killed by his captors by being kept awake. They took it in turns to watch him and kept him from rest so that 'at last he was quite wearied out and died'.*

Research in animals has shown that sleep may be as important as food for survival. A series of experiments on rats forced them to stay awake by making them move every time they began to drift off. They ate more than usual but lost weight and their general physical condition deteriorated, no matter what the researchers did to alleviate it. Only sleep helped. After two weeks of continuously being awake the rats died, and not because of weight loss or any other physical reason that could be determined. Sleep kept the rats healthy and lack of it meant they died.

Sleep allows the body to carry out running repairs. People who are sleep-deprived tend to have higher blood

---

* Sleep deprivation has been used as a method of extracting confessions even up to the twentieth century. Sleep deprivation is also one of the tactics used by Petruccio to 'tame' Kate in *The Taming of The Shrew*.

pressure and compromised immune systems, which can lead to infections. Symptoms of depression can appear, along with seizures and migraines. One theory to explain the effects on the mind is that sleep allows the brain to remove waste material from brain cells more efficiently than during wakefulness. This theory is backed up by studies in sleep-deprived animals that show changes on the brain and brain stem that vary by species. Lack of sleep certainly killed these animals, and more quickly than lack of food, but the end result might not be the same for humans.

Sleep-deprivation studies have also been carried out on people. The most famous was done in 1964 by Randy Gardner, a 16-year-old student from San Diego in California, who stayed awake for 11 days for a school science project.*

The experiment was conducted under medical supervision. His observers noted that as the experiment progressed Gardner's attention span suffered and he became sullen and irritable. More alarmingly, on the third day he mistook a street sign for a pedestrian. By the fourth day he was convinced he was a professional football player and got upset when anyone questioned his skills. By the ninth day he struggled to complete sentences and blurred vision became an increasing problem. The experiment was stopped after 264 hours because Gardner had reached his goal of beating the previous record of 260 hours. He then slept for over 14 hours and made a full recovery.

---

* The experiment has been entered into the *Guinness Book of World Records* as the longest time a human has gone without sleep, but they no longer track records like this as it is too dangerous.

No one in medical history has ever been able to deny a person sleep to the point of death.* Microsleeps, an uncontrollable period of sleep lasting between a few microseconds and ten seconds, are characterised by brief memory lapses, the head bobbing down and brief loss of muscle control. They occur most frequently when someone is tired and trying to fend off falling asleep. They happen so easily that people are often only aware they have occurred when they jerk awake again. The human need for sleep may be so strong that you could never die for lack of it, but that is not a recommendation to try and test the theory. Experiments that would answer the question conclusively cannot be carried out ethically.

Lack of sleep may not have killed Lady Macbeth but it could certainly have contributed to her death. Increasing mental deterioration, hallucinations and poor attention from sleep-deprivation can lead people to do irrational and potentially dangerous things. Shakespeare's portrayal of Lady Macbeth's decline, through extreme emotional distress and anxiety leading to sleepwalking, then insomnia and finally death, is a plausible sequence of events. The theory is supported by a later speech in the play, from Malcolm, who repeats a rumour that Lady Macbeth killed herself.

Where Shakespeare got the idea for such an elaborate series of changes leading up to death is not clear. It certainly wasn't from the history books. Macbeth, Duncan

---

* There is a very rare genetic condition, Fatal Familial Insomnia, that prevents its sufferers from sleeping; but it is believed that it is the damage done to the brain by the disease that causes death, and sleeplessness is a side effect.

and Lady Macbeth, as well as many other characters in the play, were real historical figures. But Shakespeare's version of events goes far beyond a bit of artistic licence.

In real life, Macbeth was the Mormaer, effectively Lord, Earl or even King, of Moray, and his wife, Gruoch, was also of royal descent. They had an excellent lineage and justifiable claim to the throne. They would have expected to rule after the death of Malcolm II in 1034. But instead the crown went to Malcolm's grandson, Duncan, causing a considerable amount of upset. Duncan I of Scotland was not the gentle, revered older statesman of the play. He was a rash youngster who had been promoted well beyond his level of competence. He was an unpopular king to say the least.

After six years of troubled rule, Duncan met Macbeth in battle on 10 August 1040, where Duncan was killed, but it was not necessarily Macbeth who killed him. Whatever really happened on the battlefield, Duncan was certainly not stabbed in his sleep at Macbeth's castle as is his fictional counterpart. Also, when the real-life Duncan died there was little controversy; Macbeth was immediately accepted as King of the Scots and reigned successfully for 17 years.

But Shakespeare had not simply invented the image of Macbeth as an unpopular bloodthirsty tyrant; nor had his main historical source, Holinshed's *Chronicles*. Macbeth had already been portrayed as a murderer and usurper for more than 200 years, a story started by John of Fordun in his 1380 *Chronica Gentis Scotorum*. And the play wasn't merely repeating the accepted version of events; it was written with his audience, King James, very much in mind.

The Scottish King James claimed he was descended from Banquo, Macbeth's right-hand man in the play, who Macbeth also has brutally murdered. Shakespeare included the prophecy that, though Banquo would never rule, his descendants would: 'Thou shalt get kings, though thou be none'.* The playwright also made references to the recent Gunpowder Plot against the monarch,† and added witches to the story to please the witchcraft-obsessed monarch.

But what was the truth about Lady Macbeth, or Queen Gruoch Macbeth as she would have been titled? Her history is vague to say the least, but she almost certainly wasn't the cold-hearted, power-grabbing schemer of Shakespeare's play. One of the Bard's most interesting and iconic characters seems to have sparked little interest in historians and chroniclers. Even the dates of her birth and death are not known for certain (though she probably died in 1054). Had she died of something so unusual as lack of sleep, it might have been expected to be commented on, but it seems the true cause of her death has not been recorded.

---

* Later historians have shown that Banquo never actually existed and trace James's lineage from a Breton family that relocated to Scotland after Macbeth's time.

† The plan was to blow up the Houses of Parliament on 5 November 1605 when King James was in attendance.

CHAPTER ELEVEN

Exit Pursued by a Bear

> If this were played upon a stage now, I could condemn it as an improbable fiction.
>
> *Twelfth Night*, Act 3, Scene 4

In the above quote from *Twelfth Night*, Fabian acknowledges that not everything that happens in Shakespeare's world is always realistic. This chapter follows his lead. It is all about the slightly ridiculous, perhaps unbelievable, deaths that you would have thought the product of a very fertile imagination. But sometimes fact is stranger than fiction.

Shakespeare certainly loved the fantastical and theatrical. From a writer who included fairies, living statues and a man with an ass's head in his plays, surreal and strange goings-on are to be expected, and deaths are no exception. However, what seems implausible or strange for today's theatre-goers would not always appear in the same light for Shakespeare's audiences.

* * *

On a sliding scale of silliness, the deaths of two characters burnt to a crisp by a bolt of lightning seems a pretty ludicrous end, more suited to a comedy than a tragedy. It is almost a cartoon death where a Looney Tunes character is struck by lightning and instantly reduced to a pile of ash. People certainly are killed by lightning, but it is such a rare and unusual event that it seems strange to include it in a play. In fact, Shakespeare was following the events described in the source for his play, book eight of the *Confessio Amantis* by John Gower.

In *Pericles*, the eponymous hero travels to Antioch, in the very south of modern-day Turkey, to ask King Antiochus for his daughter's hand in marriage. The King agrees, but first Pericles must solve a riddle. If he can't find the answer, the punishment is death. Pericles accepts the challenge, but it is a trap. The riddle reads,

> I am no viper, yet I feed
> On mother's flesh which did me breed.
> I sought a husband, in which labour
> I found that kindness in a father:
> He's father, son, and husband mild;
> I mother, wife, and yet his child.
> How they may be, and yet in two,
> As you will live, resolve it you.

The answer to the riddle is that the King is having an incestuous relationship with his daughter. To reveal this would also mean death for Pericles. He manages to escape with his life and goes on the run. But he doesn't have to worry long. In the second act news is brought to Pericles that Antiochus and his daughter have been killed: 'When he was seated in a chariot / Of an inestimable value, and his daughter with him, / A fire from heaven came and shrivell'd up / Their bodies, even to loathing'.

Lightning appears to have not only killed the King and his daughter but badly burnt their bodies in the process. This might be what would be expected from a lightning strike, but it is rarely what happens. It might suggest that Shakespeare was not familiar with the effects of lightning on a body and was writing what his audience would expect to hear in such cases.

Lightning storms are relatively common near the equator, but not so much in more northern climes such as England. Deaths from lightning strikes in England are consequently rare, but not unheard of. It seems unlikely that Shakespeare would have witnessed the results of a fatal lightning strike personally. However, their rarity could have increased interest when they did occur. News and gossip about such events would have spread rapidly.

Lightning has been seen as a punishment or weapon of the gods since ancient times. It is hardly surprising that this natural phenomenon could inspire such dread. The awesome sight and tremendous power of lightning storms have thrilled and terrified for millennia. In many respects it is right to be fearful.

A bolt of lightning can carry 150,000 amps, tens of millions of volts, and incredible heat (28,000°C, hotter than the surface of the sun). It is no wonder that so much energy concentrated into a ray just 2–5cm (1–2in) across can cause a lot of damage. Lightning can rip trees apart and demolish buildings. There is certainly enough destructive energy to kill a person, even several people. It can kill via a direct strike or by a 'side flash', where the lightning strikes another object and then jumps to the victim, or by conduction through an object. What is perhaps most surprising about lightning strikes is that the majority of those who are struck survive.

Pulses of lightning are incredibly short-lived, existing for only milliseconds, so there is less time for damage to occur than from, for example, touching high-voltage cables. Lightning follows the path of least resistance to the ground and our skin offers a lot of resistance. Human beings are therefore not very good conductors of electricity. Sweat or rain-soaked clothes, however, are far better at conducting electricity and offer an easier pathway for the lightning. The energy from the lightning as it passes through can superheat the water into steam, causing clothes to be ripped off as though there has been an explosion. The skin can be burned, often severely, by the steam or by the energy of the electrical current forcing its way through a resistive material.

Though burns can be fatal, this is not usually what kills in the case of a lightning strike. The real danger is if the electricity can penetrate the skin and enter the body. Wet skin, from rain or sweat, has a much lower resistance than dry skin. But once inside the body, tissues full of water and electrolytes offer very little resistance to the flow of electricity. The nervous system, which normally operates on electrical signals less than a tenth of a volt, can be thrown into chaos. Lightning will take the shortest path through the body to the ground. If that path is through the brain or heart, you are in real trouble.

Over-stimulation of the brainstem, and in particular the medullary respiratory centre of the brain that controls breathing, can kill quickly. Respiratory arrest can also occur when the passage of current through the thorax causes the intercostal muscles and diaphragm to go into spasm or become paralysed. But these are rare occurrences and, if the victim can be reached in time, breathing can be supported artificially and they may survive. The majority of deaths from lightning strikes are thought to be due to electrical stimulation of the heart causing fibrillation (very rapid beating of the heart). Without correction from a defibrillator, fibrillation can lead to cardiac arrest and death. Before effective methods of cardiopulmonary resuscitation had been developed, individuals had little chance of recovering from such effects. They were unlucky.

For every person killed by lightning there are 10 or 20 more who survive. Some emerge from the experience relatively unscathed, but for others it can cause serious injury and lasting health problems, from deteriorating sight to tinnitus, depression, dizziness and fatigue. Why individuals have such different outcomes is not known.

In the play, if the King of Antioch and his daughter were sitting in their carriage when they were hit, their heads would have been the highest point and probably where the lightning struck. Both the brain and the heart would be in the direct path as the lightning moved down through the body to the ground. Most likely they were killed very quickly.

The damage to the bodies, 'shrivell'd up', is unusual. It is well known that injury from lightning is capricious and unpredictable. Two people can stand side by side during a flash and one may be mutilated and killed while the other is unharmed. The degree of damage to tissues is proportional to the actual quantity of electricity flowing through them. Even in fatal cases, the physical damage can range from virtually nothing to gross burning. Feather, or fern-like, patterns on the skin (sometimes called Lichtenburg figures) are well known but not as common as textbooks might suggest and usually disappear after a few days. Irregular red marks may follow skin creases, especially if they are damp from sweating. Metal objects close to the skin may leave burns and blistering or charring are also present in some cases, but deep burns are relatively uncommon.

Shakespeare adds an unusual little detail about the bodies: 'for they so stunk, / That all those eyes adored them ere their fall / Scorn now their hand should give them burial'. It seems a strange thing to mention but it serves an artistic purpose, and may also show that he knew more about the effects of lightning strikes than it first appears.

The stink may well be artistic licence to highlight the corruption of the pair's sin. But there may also be some truth behind it. In cases of death by lightning there is often a smell of singeing or burning about the body and its clothing. If burning is more extensive it will be far worse. The smell of a burnt body is both difficult to describe and unmistakable. It is a combination of burnt flesh, an unforgettable and awful stench, and burnt hair, an unpleasant sulphurous smell. Many firefighters testify that once smelled, it is impossible to forget.

And it may not be just the burning that makes the bodies in *Pericles* smell so bad. The body of a man killed by lightning in May 1666 was said to give off an appalling stench when surgeons came to dissect it. And given the conditions regularly encountered in seventeenth-century dissections, the smell must have been staggeringly bad to be worthy of comment. They carried on regardless and found burning to the skin but no damage to the internal organs. Another possibility for the bad odours is illustrated by a more recent case. It was theorised that lightning had struck a man on his belt buckle, where it had entered the body and ruptured his intestines through rapid expansion of the gases inside.

In a few short lines Shakespeare could convey an incredibly dramatic event far better than if he tried to depict it on the stage. Thunder and lightning storms could be mimicked in a theatre using sound effects and pyrotechnics, but showing a lightning bolt hitting two characters onstage would be difficult. It is much easier to have it described by someone else and also gives the opportunity to go into a little gross detail. To have the shrivelled, badly burnt bodies shown onstage would require special props to be made – not impossible, but expensive and hardly worth it when the audience can produce something far more macabre in their minds from the description.

The deaths of King Antiochus and his daughter may be seen as just punishment owing to the severity of their crimes. On this occasion it was divine retribution rather than a court of law that brought about their execution. Other Shakespearean characters willingly offer themselves up for strange deaths for more noble reasons.

* * *

In *The Merchant of Venice* one character is willing to risk his life in the most unusual way in order to help a friend. Antonio offers to stand guarantor for a loan from Shylock. If the debt cannot be paid on time he agrees Shylock can take a pound of flesh from his body. He must have been aware that such a procedure could kill him. But the proposal is so outlandish, almost as unbelievable as being unable to pay the debt in time, that he agrees.

Claiming your 'pound of flesh' is a phrase that has entered into everyday usage and is commonly understood as a ruthless demand for what is rightfully yours, regardless of the consequences to the other person. It has even appeared in films and TV series as a particularly cruel way of killing someone. But where Shakespeare got his idea from is a mystery.

Where this pound of flesh is to be taken from isn't specified in the contract. If taken from a well-nourished thigh or buttock, Antonio might live, but it is implied it will be taken from the chest, near the heart. What is clear from the play is that the procedure is expected to kill Antonio. It is only at the last minute, and thanks to Portia's intervention, that he escapes having to pay his forfeit.

Surgical procedures were basic in Elizabethan England but not necessarily fatal. Limbs were removed and surgery performed on tumours or 'a disease that must be cut away' (*Coriolanus*), but surgeons were rarely brave enough to delve inside the torso. The risks from blood loss and infection were too great. One exception was Caesarean section, performed in only the most dire of circumstances, which was almost universally fatal for the mother. In *Macbeth* when Macduff 'was from his mother's womb / Untimely ripp'd' it would only have been done to save the child when there was no hope for the mother.

Regardless of the operation, the pain would have been excruciating as there were no anaesthetics available. Only basic pain relief could be obtained from opium and other plant extracts such as mandrakes (see Chapter 2).

Antonio is lucky not to have to go through with his procedure. Other characters are not so fortunate. In *Titus Andronicus*, two characters have body parts removed without the skill of a surgeon or any kind of pain relief.

⋆ ⋆ ⋆

Titus Andronicus is a revenge tragedy Shakespeare wrote with George Peele. The play takes a lot of inspiration from the Roman writer Seneca's most famous tragedy, *Thyestes*. As is often the case in revenge, it is not simply an eye for an eye. As Atreus puts it in *Thyestes*, 'You do not avenge crimes unless you surpass them'. Shakespeare and Peele's play begins with a dismemberment and doesn't let up for two blood-soaked hours.

At the start of the play Titus returns to Rome victorious from war with the Goths. He has brought with him Tamora, Queen of the Goths, and her three sons, Alarbus, Chiron and Demetrius, as captives. Despite Tamora's protests, Titus sacrifices Alarbus by having his body dismembered and burnt. It is the start of a cycle of bloody revenge between the two families.

To avenge Alarbus's death, his brothers, Chiron and Demetrius, kill Bassianus, who is betrothed to Titus's daughter, Lavinia. They then rape Lavinia and cut off her hands and tongue so she can't reveal what has happened. Lavinia's assault isn't shown. The aftermath of her walking onstage, mouth and stumps bleeding, is horrific enough.

The two brothers also frame Titus's sons, Martius and Quintus, for the murder of Bassianus. Martius and Quintus are arrested and sentenced to be executed for the murder but Titus is tricked into thinking he can save them by sending his severed hand to the emperor. Titus readily agrees to save his sons and his hand is cut off onstage with an axe.

Both Lavinia and Titus survive having parts of their bodies cut off. And though there is no mention of it, both would have needed treatment afterwards to survive. Stemming the loss of blood is vital. Immediate action could be taken with tourniquets or bandages. Wounds could be cauterised with hot irons and blood vessels could be tied up with threads. It was rudimentary but largely effective. Sophisticated stitching techniques and life-saving blood transfusions had simply not yet been invented. No one would have even considered sterilising anything before proceeding. It is a wonder they, or anyone else operated on before the nineteenth century, survived. The risks were well known, as Menenius hints at in *Coriolanus*, 'a limb that has but a disease; / Mortal, to cut it off; to cure it, easy.'

What is worse, Titus cuts off his hand for nothing. A messenger soon arrives returning his hand along with the severed heads of his sons. The body parts are collected up by the survivors; Titus takes one head, Marcus another and Lavinia carries away her father's hand in her mouth.

All of this is just a warm-up for the really gory bit. Every body part that has so far been lopped off is but a teaser before the main event. Titus now seeks revenge for the rape of his daughter and deaths of his sons.

First he meets with the culprits, Chiron and Demetrius, and has them bound and gagged so he can confront them with the list of their crimes. Then, like a true stage villain, he spells out what he is going to do to them: 'This one hand yet is left to cut your throats, / Whilst that Lavinia 'tween her stumps doth hold / The basin that receives your guilty blood.'

The basin is a wise precaution. Cutting the throat will result in massive blood loss as the neck contains some very large blood vessels. Closest to the surface are the jugular veins, which drain blood from the brain to the heart. Deeper within the neck are the carotid arteries, which normally take oxygenated blood up from the heart to the brain. Blood loss from arteries is more rapid than from an equivalent-sized vein because the blood is under pressure. Even if the knife doesn't cut deep enough to hit the arteries, blood loss will be significant. Unconsciousness will be rapid, within minutes, followed by death soon after.

All of this happens onstage and seeing it acted out can be quite the spectacle – some productions have caused multiple faintings. Shakespeare's audiences would have been more hardened to such sights because of their familiarity with the slaughterhouses dotted around the city and with public executions. Getting hold of large volumes of blood would not have been difficult but controlling it onstage would have been important. Scenes followed on one from another without a pause – there would have been no opportunity to clear up. Lavinia's bowl is needed to catch any blood to stop actors slipping and sliding in the mess in subsequent scenes.

However, it doesn't have to be that gory. Just seeing Titus with the knife and Lavinia with the bowl, we know that this is going to be very bloody – we don't have to see it. The actors playing Chiron and Demetrius can be turned away to shield the audience's eyes from the worst. In an Elizabethan theatre this was more difficult, as spectators were allowed on the galleried area above the back of the stage at the Globe and on the stage itself at the Blackfriars Theatre. Alternatively, depictions of the mutilations and deaths don't have to be realistic to have an impact. Several modern productions have used red ribbons in place of blood to astonishingly good effect.

But in the play Titus isn't finished yet. What he does next is even more distasteful to modern eyes – 'Hark, villains! I will grind your bones to dust / And with your blood and it I'll make a paste, / And of the paste a coffin I will rear / And make two pasties of your shameful heads'.

Chiron and Demetrius's remains are baked into a pie and served to their mother at a banquet. The idea probably came from Shakespeare's source, *Thyestes*, which sees the eponymous hero seducing his brother Atreus's wife and stealing his kingdom. Atreus gets his own back by tricking Thyestes into eating a banquet prepared from the flesh of his own sons.

Cannibalism is considered taboo, but there have been exceptions.* Up until the nineteenth century body parts were often used in medicine. Blood, extracts of Egyptian mummies and other body bits were swallowed or worn as talismans against ill health – to no physical benefit, it might be added. Although people weren't serving up human flesh at their sixteenth-century dinner table, this is not true of all cultures in all times.

Evidence from crushed skulls suggests that humans living half a million years ago ate human brains, and the practice of eating humans has continued among various peoples until very recently.† Cannibalism can be seen as an act of remembrance, combining the bodies of dead relatives with the living, or an act of dominance. Eating the heart of an enemy shows complete power over them. Others have been forced into cannibalism because of circumstances. Those stranded at sea or isolated in remote parts of the world have often turned to each other for food, some more reluctantly than others.

Starvation or revenge can be a powerful motivator. Eating human flesh because of a preference or craving for the taste is very rare indeed. Such flesh usually comes from murder victims. Peter Stumpp was one such person

---

* One possible current exception in the West is eating the placenta after a birth, which is thought to ward off post-partem depression, though women are usually eating their own. Recipes are available online and have even been included on a TV cooking show in the UK in 1998 – though it received several complaints and a rebuke from the Broadcasting Standards Commission (now Ofcom).

† The Korowai in Indonesian New Guinea are thought to be the only group of people still practising cannibalism, but they do not see it as eating humans. They eat *khakhua*, beings who come disguised as a relative or friend of a person they want to kill.

who had a taste for human flesh. When he was captured near Cologne, Germany, in 1589, he confessed to having killed and eaten at least 16 people. He was executed as a werewolf.

Another tale from Shakespeare's day is that of Alexander 'Sawney' Bean, the head of a Scottish clan said to have killed and eaten over 1,000 people. Sawney and his wife, Agnes Douglas, who was also said to be a witch, are supposed to have started their cannibalistic life when they attacked and killed a passer-by. It became too risky to sell on the victim's valuables for food so instead they ate him. The couple withdrew to live a reclusive life in a cave on the east coast of Galloway where they raised a large family of cannibals. Their home was so hidden that few locals seem to have realised the family was living there. They also apparently failed to notice the disappearance of hundreds of people, as the family needed to feed the parents, 14 children and 32 grandchildren. When they had had their fill, the remaining flesh was salted, dried and pickled. They were only discovered when one of their victims escaped.

Though the tale of the Sawney clan would have been a rich source of inspiration for Shakespeare, there is considerable doubt that they ever existed. The story first appeared in *The Newgate Calendar* in the nineteenth century. Sixteenth-century ballads and broadsheets fail to mention the family or the disappearance of their victims.

Even if he was a dedicated meat eater, there were presumably alternative food sources available to Peter Stumpp, and even the Sawney clan could have presumably fed on game and wildfowl. Can the taste of human flesh really be so wonderful that it would drive people to kill for it? Some have claimed that human flesh has a particularly desirable flavour. In the 1880s, Alfred Packer, a prospector who turned to the bodies of his five companions for food when provisions ran out, told a reporter that the breasts of men were 'the sweetest meat' he had ever tasted.

The flavour of any meat does of course depend on where the cuts are taken from and how they are prepared. One commonly known cannibal myth is that human meat tastes like pork, hence the name 'long pigs' for cannibals' victims. However, in 1878, a sailor on a drifting schooner described the flesh of one of the dead crewmen as being 'as good as any beefsteak' he ever ate. William Seabrook, the twentieth-century occultist, traveller and cannibal, agreed. He had written about cannibalism from his travels in West Africa and said the meat he saw looked like beef, but he later admitted the cannibals had not let him partake in their rituals. Instead he travelled to Paris and with the help of a hospital he acquired a chunk of human meat from a recently deceased person. He cooked and prepared it and reported that, 'in colour, texture, smell as well as taste … veal is the one meat to which this meat is accurately comparable.' And according to the anthropologist Stanley Garn, the lean–fat ratio of human flesh does indeed make it similar to veal.

The flavour and appearance might vary but perhaps it wouldn't taste so very different to other meats commonly eaten. In *Titus Andronicus*, there are even hints of a recipe: 'Receive the blood: and when that they are dead, / Let me go grind their bones to powder small / And with this hateful liquor temper it; / And in that paste let their vile heads be baked.' It seems he is preparing a kind of blood pastry to cover the meat. Tamora is unlikely to have been alerted to unusual ingredients in the pie from the taste and Titus gets his revenge.

Violent extremes are more associated with sudden flares of temper and lashing out in the heat of the moment. But Titus's plans are cold and calculating. He has his family with him to support him and conspire with him. Tamora also has her lover Aaron plotting with her and manipulating events. *Titus Andronicus* shows how groups of people can egg each other on to ever more violent extremes. Another play, *Julius Caesar*, also has groups of people coercing each other into ever greater acts of violence.

⋆ ⋆ ⋆

The group of conspirators that killed Caesar would be unlikely to act in the way they do as individuals, but together eight of them inflict 33 stab wounds on the Emperor. One vicious attack leads to another. The citizens of Rome are enraged by Caesar's murder. When a group of people come across a man called Cinna, they immediately assume he is the Cinna who took part in Caesar's murder. They want revenge and of course it must be a more bloody murder than Caesar's. The mob doesn't even care that they have the wrong Cinna. One citizen shouts, 'Tear him, tear him!' The man desperately tries to explain, 'I am Cinna the poet, I am Cinna the poet', but the crowd is out for blood: 'It is no matter, his name's Cinna; pluck but his name out of his heart, and turn him going.' He is literally torn apart.

Mob violence can undoubtedly result in fatalities but pulling a body apart is difficult. We know this because in medieval times people were torn apart as a form of execution – and it required the help of horses. Even then it was not always enough and an axe was needed. The death of Cinna the poet may seem improbable but Shakespeare hadn't invented this macabre death for artistic reasons; he was depicting real events from history.

From the audience's point of view, seeing a man being torn apart is potentially even more unpleasant than the horrific goings-on in *Titus Andronicus*, but it probably wouldn't worry an Elizabethan crowd used to seeing dogs and bears torn apart for sport. It would need a lot of blood, flesh and props to look convincing and would create one hell of a mess to clear up before the next scene. Shakespeare spares everyone the trouble and has Cinna bundled offstage to his fate, giving the audience the opportunity to imagine far worse than could ever be depicted onstage.

Seeing fake limbs and blood being thrown about could also easily descend from the macabre to the ridiculous and invite laughter from the audience – not the mood you are looking for in a tragedy. Such a reaction might not be all that unusual from an Elizabethan audience that watched animals fight to the death for entertainment. Blood sports were hugely popular at the time and among every class, from the poorest even to royalty. These events have been described by recent critics as a 'carnival of cruelty' where 'again and again the audience was pleased by what it saw, cheered it on and laughed at it'. A whole menagerie of animals were set against each other or beaten in public, but the most popular by far was bear- and bull-baiting.

★ ★ ★

Bear-baiting and theatre were closely linked in Elizabethan London. They often shared the same venues and one playhouse was specifically designed with a removable stage to make way for the animals on the days plays weren't performed.* Playwrights frequently borrowed the language and style of bear-baiting events for their dramas. Cinna the poet is just one example of a character put in the place of a bear trapped among a vicious pack about to tear him apart. Ben Jonson took advantage of the bear-baiting that also took place in the Hope Theatre to add an extra dimension to his play *Bartholomew Fair*.

Bartholomew Fair was a real annual event that was put on every summer from 1133 until 1855, when it was suppressed for debauchery. It took place over several days at Smithfield, to the north-west of the city walls. The fair was originally for cloth-trading but by Jonson's day there was much more on offer. Smithfield had by then established itself as a livestock market and was also known as the site of many executions. At the fair, among the puppet shows and food stalls there were also acrobats and wild animals on display. At the Hope Theatre, food sellers

* Of all the London theatres in the late sixteenth century the Globe was special, not just because it was the home of Shakespeare's theatrical company, but because it was the only theatre to be used exclusively for the performance of plays.

walked through the crowd just as they would at the real fair, and the venue would retain the animal stench from the bear-baiting events of the previous day, adding considerably to the atmosphere of the performance.

In Britain bears were hunted to extinction around 500 AD, but they were a common sight on the streets of sixteenth-century London thanks to imports from continental Europe. The bears were made to perform on street corners so they could be laughed at by passers-by. They were considered supremely ugly animals and so to dress them up and make them dance was hilarious to the average Elizabethan. What was even more fun was to see them physically attacked with whips and dogs.*

A bear would be chained to a post and English mastiffs would be set on him. The bear would fend off the attacks, severely injuring the dogs, but they would only return to attack the bear again until they were killed. Sometimes the owners would intervene before a dog was fatally injured but sometimes it happened so quickly that fresh dogs had to be brought out to continue the entertainment lest the audience feel they had been short-changed. The bears, however, were expensive commodities and owners went to some lengths to look after them and patch them up again ready for another fight.

Part of the attraction of these blood sports was what it revealed about the animals. English mastiffs were highly regarded because they would never give up the fight. Bulls that fought off the attacks from dogs with their horns were admired for their cunning. Bears were seen as

* Bear-baiting also happened on the continent but it was considered an English speciality.

'artfully' keeping the dogs at bay. Other combinations of animals were tried with varying degrees of success. Horses with a monkey on their backs were popular because the monkey shrieked when it was bitten by the dogs that chased it. Lions on the other hand were a disappointment. Despite their fierce reputation, when faced with a pack of dogs they retreated to their den and refused to fight.

Blood sports weren't popular with everyone, however. Puritans disliked bear-baiting because it was carried out on a Sunday. Others found the whole spectacle distasteful and risky. Thomas Nashe in his *Anatomie of Abuses* describes the sport: 'besides that it is a filthy, stinking and loathsome game, is it not a dangerous and perilous exercise wherein a man is in danger of his life every minute of an hour?' The animals certainly presented a risk but Nashe may also have been referring to the dangers from the shoddy construction of the arenas where bear-baitings were held (see Chapter 2).

Shakespeare also took advantage of the audience's familiarity with the violence associated with bears and bear-baiting. He referenced the sport to highlight underlying violence or threat in parts of his plays. For example, in *Macbeth*: 'They have tied me to a stake; I cannot fly, / But, bear-like, I must fight the course.' The use of bear-baiting references in plays also reveals a lot about the characters. *Twelfth Night* contains numerous references to bear-baiting. Sir Andrew Aguecheek is a keen fan; he, Sir Toby and Fabian all refer to the blood-sport directly. The name 'Duke Orsino' is a play on the word 'ursine' meaning bear-like (from the Latin for bear, *ursus*). Furthermore, Malvolio's treatment at the hands of Sir Toby and his coevals is framed as a bear being baited.

He is placed in a dark room and taunted. At the very end of the play Malvolio promises to 'be revenged on the whole pack of you'. Similar behaviour is acted out towards Gloucester when he is blinded in *King Lear*, perhaps in reference to the popular practice of taunting blinded bears with whips.

Some bears became well-known local 'characters'. They were given names and acquired personalities. There was Ned Whiting, George Store and Harry Hunks (one of the blinded bears that was tormented with whips). One female bear, known as Old Nell of Middlewich, was taken into ale houses for a drink. But perhaps the most famous of all the bears was Sackerson, immortalised by his inclusion in a Shakespeare play. In *The Merry Wives of Windsor*, Slender brags of his encounters with the famous bear: 'I have seen Sackerson loose twenty times, and have taken him by the chain; but, I warrant you, the women have so cried and shrieked at it, that it passed: but women, indeed, cannot abide 'em; they are very ill-favoured rough things.'

Bears might have been hilarious fun when they were chained to a post but the mood rapidly changed if they got loose, as they evidently did on occasion – as Fabian comments on Cesario's fear in *Twelfth Night*, 'he pants and looks pale, as if a bear were at his heels'.

When a bear suddenly appears onstage in *The Winter's Tale* it was almost certainly less incongruous to Shakespeare's audience than it seems today. The moment in the middle of Act 3, Scene 3 is perhaps Shakespeare's most famous stage direction [*Exit, pursued by a bear*]. The character being pursued is Antigonus, who has just landed on the coast of Bohemia.

Shakespeare's acting company could presumably have borrowed a bear. Even though their home theatre, the Globe, did not host blood sports, there were at least five bear-baiting arenas nearby. But the actors who would have to share the stage with the animal might not have been so enthusiastic. There might also have been considerable enthusiasm among the cast to play the bear themselves. There is no surviving list of props for Shakespeare's company but their rivals, the Lord Admiral's Men, are known to have had 'i bears head' and 'i bears skin' as well as various other animal parts, including heads of a Cerberus, a bull's head, one lion's skin and two lion's heads.

How realistic an actor in a bear costume would be is hardly relevant. Factual details and realism have nothing to do with *The Winter's Tale*. The title of the play indicates that this is a story and any facts that might work their way into the plot are just a bonus. For example, Bohemia, a region roughly equating to the modern-day Czech Republic, might not have a coast for Antigonus to land on but it does have bears to chase him.

The Eurasian brown bear is common in central Europe but in its natural habitat it is relatively shy of humans. Encounters with these bears are rare and attacks on humans rarer still. When they do occur it is usually because a bear, often a female with her cubs, has been disturbed. Shakespeare's experience of bears would have been very different. Bears chained to posts and tormented and abused regularly are likely to act differently from those left to their own devices in the wild.

Clearly Antigonus disturbed the bear, which let out 'A savage clamour!' He then does the worst thing possible and runs. Running away from a bear is only likely to encourage it to follow, exactly as it does in *The Winter's Tale*. The chase is the last we see of both the bear and Antigonus. What happens next is reported by a clown who happened to witness the attack and sensibly kept out of it.

The clown tells 'how the poor gentleman roared and the bear mocked him, both roaring louder than the sea or weather', and 'how he cried to me for help and said his name was Antigonus, a nobleman'. The fight between the two is an unfair one. Eurasian brown bears are much bigger and heavier than humans and come equipped with 42 teeth, including some very big ones for biting prey, and claws that can grow up to 10 centimetres (4in) long. It is no surprise that 'the bear tore out his shoulder-bone'.

The few bear attacks on humans that occur in Europe today are not usually fatal. Croatia saw its last bear-related death over 65 years ago. Sweden has gone over a century without having a human killed by a bear. While the damage to Antigonus's shoulder might not be fatal today, it almost certainly would have been over 400 years ago. And the attack doesn't stop there: 'the bear half dined on the gentleman'. The clown leaves the animal to its dinner and only later returns to 'see if the bear be gone from the gentleman and how much he hath eaten: they are never curst but when they are hungry: if there be any of him left, I'll bury it.'

Humans are an unusual meal for wild Eurasian brown bears. Their diet has changed dramatically over time. In ancient times their diet was approximately 80 per cent carnivorous, but by the Middle Ages this had diminished to 40 per cent. Today it is around 5–10 per cent and their preferred meat is either found carcasses of animals that have died a natural death or animals that are very easy to kill, mostly sheep.

Several Shakespearean deaths may seem absurd or unrealistic to modern audiences, but he was writing in a world very different from the one we live in now. People eaten by bears or baked into pies would not have raised nearly so many eyebrows when they were first shown onstage in Renaissance London.

Epilogue

Good plays prove the better by the help of good epilogues.

As You Like It, Act 5, Scene 4

Shakespeare's work is not always easy to categorise. His tragedies have funny moments and his comedies include deaths; 'The web of our life is of a mingled yarn, good and ill together' (*All's Well That Ends Well*). As this book shows, the deaths are just as difficult to file under nice neat headings as the plays he portrayed them in.

The Bard killed off over 250 named characters in his plays and poems, in dozens of different ways. In many respects he reflected the time he was writing in. He lived in an unenviable era as far as death and violence were concerned. Unlike today, bears, sword fights and executions could be responsible for the untimely end of Londoners in Elizabethan and Jacobean times. In other ways the reality of life and death in Renaissance England is poorly reflected in his literary output. Plague and the majority of contagious diseases are all but ignored. But Shakespeare was not documenting the life and death he saw around him. He wrote to entertain the thousands of people who packed themselves into the playhouses day in and day out.

Every death included in his work, and the manner of it, serves a dramatic purpose. Some deaths show the injustices of life and others bring justice to those who have done wrong. There are accidents, planned deaths and unavoidable deaths. Some are passed over quickly, while others are described in detail. His first loyalty was to the drama itself; historical accuracy and realism were only secondary considerations. The comedies, in places, stretch credibility, but every fantastical moment adds to the enjoyment of the play. His histories play havoc with timelines and motives, but all the essential historical elements are still there. The tragedies may seem bloody and violent, but they are no more so than modern TV series. And, among all the drama and theatricality, there are some extraordinary insights into the processes of death. The accuracy of some of the minutiae that he included suggests Shakespeare had observed them first-hand.

Death is a part of life that today is rarely witnessed outside of sanitised and controlled situations, which means that unlike our predecessors we are not always familiar with the reality of dying. It is one of the many things in his plays that can be difficult to relate to for modern audiences. Over the intervening four centuries there has been a dramatic shift in attitudes to life and death, but there appears to be no lessening of enthusiasm for the Bard and his work. We may not laugh at all the same jokes, or fully appreciate comments on contemporary events, but Shakespeare still entertains and inspires like no other playwright. His impact is felt far beyond the Renaissance London stage he was writing for.

We still speak his words in everyday conversation, often without realising. Phrases such as 'heart of gold', 'break the ice', 'wild goose chase', 'seen better days' and many more were given to us by Shakespeare. He is credited with adding 1,700 words to the English language, though not all have survived the passage of time. He transformed nouns into verbs, verbs into adjectives and created other words apparently from scratch; *Hamlet* alone contains 600 words previously unknown in the English language.

Beyond his contribution to the spoken word, the breadth of his cultural appeal and influence is astonishing. His works are only outsold by the Bible and he is the third most translated author.* His plays have been staged all around the globe, from famous theatres to village halls. There have been productions in the open air and on the open sea. His characters have been dressed in everything from elaborate Elizabethan costumes to modern-day military fatigues. Over 400 operas have been derived from his work (according to *Grove's Dictionary of Music and Musicians*), and the British Universities Film and Video Council's database lists over 600 entries for *Macbeth* alone. His work has been adapted and reinterpreted for musicals, ballet, TV, books, games and memes. There seems to be no end to the creativity Shakespeare can inspire.

The richness of the writing he left us has been the source material for tens of thousands of books and articles examining his life and work, an outpouring of research and interest that shows no signs of stopping. Academics across a range of disciplines have found something of relevance to their field or a new perspective in Shakespeare's work, and death is no exception.

---

* According to an ever-fluctuating list, the Bible is the most translated work; Agatha Christie's novels take second place, with Shakespeare roughly equal third with Jules Verne.

This book has looked at only one tiny aspect of Shakespeare's work. There is so much more to explore but we are out of time. I am no good at endings so I shall leave it to the master.

> If we shadows have offended,
> Think but this, and all is mended,
> That you have but slumber'd here
> While these visions did appear.
> And this weak and idle theme,
> No more yielding but a dream,
> Gentles, do not reprehend:
> if you pardon, we will mend:
> And, as I am an honest Puck,
> If we have unearned luck
> Now to 'scape the serpent's tongue,
> We will make amends ere long;
> Else the Puck a liar call;
> So, good night unto you all.
> Give me your hands, if we be friends,
> And Robin shall restore amends.

Appendix

The summary of all our griefs

Henry IV, Part II, Act 4, Scene 1

COMEDIES

The Tempest ***(1611)***		
Many sailors die in a shipwreck but all are restored to life. Sycorax the witch, Caliban's mother, dies before the start of the play.		
Two Gentlemen of Verona ***(1594)***		
No one dies. One outlaw has been banished for stabbing a man in the heart. Proteus says he loved a woman but she died; he also thinks, erroneously, that Valentine has died.		
The Merry Wives of Windsor ***(1600)***		
No one dies but Dr Caius threatens to cut Slender's throat (he doesn't).		
Measure for Measure ***(1604)***		
The play may have been revised by Thomas Middleton after Shakespeare's death.		
Ragozine	Prison fever	Only appears on stage as a decapitated head
Lucio	Hanged	The play ends before the sentence is carried out
Claudio and Angelo		Threatened with execution but pardoned
The Comedy of Errors ***(1589)***		
No one dies but Aegeon spends the entire play under threat of execution		
Much Ado about Nothing ***(1598)***		
No one dies but at the start of the play the men have just returned from battle. Hero is believed to have died of shame (she hasn't).		
Love's Labour's Lost ***(1594)***		
King of France	Natural causes	

<table>
<tr><th colspan="3">A Midsummer Night's Dream (1595)</th></tr>
<tr><td>Mother of changeling boy</td><td>Childbirth</td><td>Does not appear on stage</td></tr>
<tr><td>Pyramus</td><td>Stabs himself</td><td>Play within the play</td></tr>
<tr><td>Thisbe</td><td>Stabs herself</td><td>Play within the play</td></tr>
<tr><th colspan="3">The Merchant of Venice (1596)</th></tr>
<tr><td>Portia's father</td><td></td><td>Dead before the play begins</td></tr>
<tr><td>Antonio</td><td>Removing a pound of flesh</td><td>The penalty is not carried out</td></tr>
<tr><td>Shylock</td><td></td><td>Threatened with death unless he converts to Christianity</td></tr>
<tr><th colspan="3">As You Like It (1599)</th></tr>
<tr><td colspan="3">No one dies but a wrestling contestant is expected to die because of crushed ribs. Oliver is almost killed by a lion.</td></tr>
<tr><th colspan="3">The Taming of the Shrew (1593)</th></tr>
<tr><td colspan="3">No one dies during the play but Lucentio is going about in disguise because he killed a man. Petruchio is recently bereaved.</td></tr>
<tr><th colspan="3">All's Well That Ends Well (1602)</th></tr>
<tr><td colspan="3">No one dies but at the opening of the play the Countess and Bertrum are in mourning for the Count. Helena's father has died before the start of the play. The King is close to death before Helena cures him.</td></tr>
<tr><th colspan="3">Twelfth Night, Or What You Will (1599)</th></tr>
<tr><td colspan="3">No one dies but Viola believes her brother Sebastian is dead and Sebastian thinks Viola is dead. Lady Olivia is in mourning for her father and brother. Orsino threatens to kill Cesario and Olivia (he doesn't).</td></tr>
<tr><th colspan="3">The Winter's Tale (1610)</th></tr>
<tr><td colspan="3">Several characters are threatened with execution (by burning, hanging, flaying, being covered in honey and sat on a wasps' nest then pestered to death by flies). Camillos is told to poison Polixenes (he doesn't).</td></tr>
<tr><td>Mamillius</td><td>Grief</td><td></td></tr>
<tr><td>Hermione</td><td>Grief</td><td>A statue of Hermione is brought to life</td></tr>
<tr><td>Antigonus</td><td>Bear attack</td><td></td></tr>
</table>

HISTORIES

Does not include the many thousands that died in numerous wars.		
The Life and Death of King John ***(1596)***		
Duke of Austria	Beheaded	
Count Melun	Slain	
Arthur	Fall from a height	
Lady Constance	Frenzy	
Queen Elinor	Dies	No cause of death given
Peter of Pomfret (a prophet)	Hanged	On the day he prophesied the King would die
Monk	Poisoned	He does not appear on stage
King John	Poisoned	By a monk
The Life & Death of Richard II ***(1595)***		
John of Gaunt	Sickness	
Duchess of York	Dies	
Bushy and Green	Beheaded	
Earl of Wiltshire	Beheaded	
Duke of Gloucester	Murdered	
Thomas Mowbray	Dies	
Two of Exton's accomplices	Slain	
King Richard	Slain	
Oxford, Salisbury, Blunt and Kent	Beheaded	
Brocas and Sir Bennet Seely	Beheaded	
Abbot of Westminster	Guilty conscience	
The First Part of King Henry the Fourth ***(1597)***		
Mortimer	Slain	
Shirley, Stafford and Blunt	Slain	Mistaken for Henry IV
Hotspur (Henry Percy)	Slain	

Earl of Worcester	Sentenced to death	Beheaded
Vernon	Sentenced to death	Hanged, drawn and quartered
The Second Part of King Henry the Fourth *(1597)*		
Double	Dies	Does not appear on stage
Sir John Colville	Sentenced to death	
Henry IV	Natural causes	
The Life of King Henry the Fifth *(1598)*		
Falstaff	A burning quotidian tertian	Does not appear on stage
Earl of Cambridge, Lord Scroop and Sir Thomas Grey	Sentenced to death	Beheaded
Bardolph	Hanged	
Nym	Hanged	
Doll	Malady of France	Syphilis
Constable of France, Lord Rambures, Lord Grandpre, Duke of Orléans, Duke of Bourbon and several other French noblemen who don't appear in the play	Slain	
Earl of Suffolk, Sir Richard Ketley and Davy Gam	Slain	
Duke of York	Slain	Or maybe crushed, or maybe heart attack

The First Part of King Henry the Sixth *(1591)*		
Possibly written in collaboration with Christopher Marlowe and Thomas Nashe.		
Earl of Salisbury and Sir Thomas Gargrave	Cannon ball	
Edmund Mortimer	Dies	After long imprisonment
Earl of Cambridge	Executed	Beheaded
Duke of Bedford	Slain	
Lord Talbot and his son John	Slain	
Joan of Arc	Burnt at the stake	
The Second Part of King Henry the Sixth *(1590)*		
Bolingbroke	Sentenced to death	Hanged, drawn and quartered
Father John Hume	Sentenced to death	Hanged
Father John Southwell	Sentenced to death	Died before his execution
Margaret Jourdain	Burnt at the stake	
Thomas Horner	Slain	Died in an 'ordeal by battle'
Duke Humphrey	Murdered	Strangled, or suffocated, but probably natural causes
Cardinal Beaufort	Guilt over an evil life	Probably natural causes
Suffolk	Beheaded	By pirates
Sir Humphrey Stafford and his brother William	Slain	By Jack Cade's mob
Clerk of Chatham	Executed	Hanged
Soldier	Killed	For calling Jack Cade by the wrong name
Lord Say	Beheaded	

Sir James Cromer	Beheaded	
Matthew Goffe	Slain	He dies on stage but doesn't get to say a line
Jack Cade	Slain	Whilst resisting arrest
Lord Clifford	Slain	
Earl of Somerset	Slain	
The Third Part of King Henry the Sixth *(1590)*		
Earl of Rutland	Stabbed	
Duke of York	Stabbed	
Father	Slain	Killed by his son
Son	Slain	Killed by his father
Lord Clifford	Slain	
Earl of Warwick	Slain	
Marquess of Montague	Slain	
Duke of Somerset	Beheaded	
Prince Edward	Stabbed	
Henry VI	Stabbed	
The Life & Death of Richard the Third *(1592)*		
Clarence	Stabbed	Then drowned in a butt of malmsey
Edward IV	Natural causes	Over indulgence
Lord Hastings	Beheaded	
Lord Rivers, Lord Grey and Sir Thomas Vaughan	Beheaded	
Princes	Smothered	Maybe
Lady Anne	Poisoned	But in actual fact natural causes
Duke of Buckingham	Beheaded	
Duke of Norfolk, Lord Ferris, Sir Robert Brackenbury and Sir William Brandon	Slain	
Richard III	Slain	

The Life of King Henry the Eighth, or All Is True *(1612)*		
Written with John Fletcher		
Duke of Buckingham	Beheaded	
Doctor Pace	Madness	
Cardinal Wolsey	Natural causes	
Queen Katherine	Natural causes	

TRAGEDIES

Does not include the nameless thousands that died in the many wars included in the tragedies.		
Troilus and Cressida ***(1601)***		
Polyxenes, Epistrophus, Cedius	Slain	
Amphimachus and Thoas	Deadly hurt	
Patroclus	Slain	
Hector	Slain	
Pandarus	Syphilis	Dies after the play has ended
The Tragedy of Coriolanus ***(1607)***		
Coriolanus	Stabbed	
Titus Andronicus ***(1593)***		
Written with George Peele		
Twenty-one sons of Titus	Slain	
Alarbus	Dismembered	Entrails then burnt
Mutius	Stabbed	
Bassanius	Stabbed	
Martius	Beheaded	
Quintus	Beheaded	
A fly	Stabbed	
Nurse	Stabbed	

Chiron and Demetrius	Throat cut	Then baked into a pie
Lavinia	Stabbed	
Tamora	Stabbed	
Saturninus	Stabbed	
Titus	Stabbed	
Aaron	Buried chest-deep and starved	
Clown	Hanged	
Romeo and Juliet *(1594)*		
Mercutio	Slain	
Tybalt	Slain	
Paris	Slain	
Romeo	Poison	A message is delayed from being delivered to Romeo because of plague
Juliet	Stabs herself	After drinking a potion that makes her appear dead
Lady Montague	Grief	
Timon of Athens *(1597)*		
Probably written with Thomas Middleton		
Timon	Dies	
The Life and Death of Julius Caesar *(1599)*		
Julius Caesar	Stabbed	33 times
Cinna the poet	Torn apart	
Portia	Swallowed hot coals	Probably carbon monoxide poisoning
Cicero	Executed	
Cassius	Falls on his sword	
Titinius	Falls on his sword	
Cato	Falls on his sword	

Brutus	Falls on his sword	
The Tragedy of Macbeth ***(1605)***		
The play may have been revised by Thomas Middleton after Shakespeare's death.		
Macdonald	Slain	He does not appear on stage
Duncan	Stabbed	
Duncan's two guards	Killed	
Banquo	Slain	
Young Macduff and Lady Macduff	Slain	
Lady Macbeth	Suicide	Maybe
Macbeth	Slain	
The Tragedy of Hamlet ***(1600)***		
Old Fortinbras	Slain	Does not appear on stage
Old Hamlet	Poison	In the ear
Gonzago	Poisoned	Play within the play
Polonius	Stabbed	
Ophelia	Drowned	
Yorick		Only appears as a skull
Rosencrantz and Guildenstern	Put to death	
Gertrude	Poisoned drink	
Laertes	Stabbed with a poisoned sword	
Claudius	Stabbed with a poisoned sword	
Hamlet	Stabbed with a poisoned sword	

King Lear *(1605)*		
Servant	Stabbed	
Duke of Cornwall	Slain	
Oswald	Slain	
Earl of Gloucester	Joy	
Regan	Poisoned	
Goneril	Stabs herself	
Edmund	Slain	
Cordelia	Killed	She is strangled or smothered then hanged
Lear	Grief	
Othello, the Moor of Venice *(1604)*		
Rodrigo	Slain	
Desdemona	Smothered	
Brabantio	Grief	
Emilia	Stabbed	
Othello	Stabs himself	
Antony and Cleopatra *(1606)*		
Fulvia	Natural causes	Does not appear on stage
Pacorus	Slain	He dies on stage but doesn't get to say a line
Marcus Crassus	Slain	Does not appear on stage
Pompey the Great	Murdered	His death is talked about but he does not appear on stage
Domitius Enobarbus	Dies	
Eros	Falls on his sword	
Antony	Falls on his sword	
Iras	Dies	After kissing Cleopatra
Cleopatra	Bite of an asp	
Charmain	Bite of an asp	

Cymbeline King of Britain ***(1609)***		
Imogen	Poison	She appears dead but revives
Cloten	Beheaded	
Leonatus	Grief	
Postumus's mother	Childbrith	
The Queen	Frenzy	

PLAYS NOT INCLUDED IN THE FIRST FOLIO

These plays are generally attributed at least in part to Shakespeare		
Pericles, Prince of Tyre ***(1608)***		
written with George Wilkins		
Antiochus orders Pericles to be poisoned but he escapes. Pericles is also the only survivor of a shipwreck. Thaisa is believed to have died in childbirth but is later revived.		
Antiochus and his daughter	Lightning	
Lychorida	Dies	No cause of death given
Transylvanian	Dies	Possibly of venereal disease
Simonides	Dies	No cause of death given
Two Noble Kinsmen ***(1613)***		
Written with John Fletcher		
Three Queens petition Theseus to recover the bodies of their husbands from the battlefield. The Jailer is threatened with hanging for helping Arcite escape.		
Arcite	Thrown from horse	
The Reign of King Edward III ***(1592)***		
Written with Thomas Kyd		
King of Bohemia	Slain	
Sir Charles of Blois	Slain	
Prince Edward	Slain	

POEMS

<table>
<tr><td colspan="3">Venus and Adonis (1593)</td></tr>
<tr><td>Adonis</td><td>Gored by a boar</td><td></td></tr>
<tr><td colspan="3">The Rape of Lucrece (1594)</td></tr>
<tr><td>Lucrece</td><td>Stabs herself</td><td></td></tr>
<tr><td colspan="3">A Lover's Complaint (1609)</td></tr>
<tr><td colspan="3">No mention of death</td></tr>
<tr><td colspan="3">The Phoenix and the Turtle (1601)</td></tr>
<tr><td colspan="3">Metaphysical poem about the death of love</td></tr>
<tr><td colspan="3">Sonnets</td></tr>
<tr><td colspan="3">There are no specific deaths but there are many references to death in general</td></tr>
</table>

Bibliography

O Lord, sir, I'll be sworn upon all the books in England

Henry IV Part I, Act 2, Scene 4

Books and Articles

Ackroyd, P. 2005. *Shakespeare: The Biography*. Chatto & Windus, London.

Adamis, D., Treloar, A., Martin, F. C., MacDonald, A. J. D. 2007. A Brief Review of the History of Delirium as a Mental Disorder. *History of Psychiatry* 18(4): 459–469.

Aristophanes. 1998 edn. *Plays: One*. Methuen Drama, London.

Bamford, J. 1603. *A Short Dialogue Concerning the Plague's Infection*. Richard Boyle, London.

Barrett, A., Harrison, C. (eds). 1999. *Crime and Punishment in England: A Sourcebook*. UCL Press, London.

Beier, L. M. 1987. *Sufferers and Healers: The Experience of Illness in Seventeenth-Century England*. Routledge & Kegan Paul, London and New York.

Bergeron, D. M. 1978. The Wax Figures in *The Duchess of Malfi*. *Studies in English Literature, 1500–1900*, 18(2): 331–339.

Boccaccio, G. 1820 edn. *The Decameron, or Ten Days' Entertainment*. R. Priestley and W. Clarke, London.

Brown, R. (ed.). 1871. Venice: April 1531, *Calendar of State Papers Relating To English Affairs in the Archives of Venice, Volume 4, 1527–1533*, 278–281. Her Majesty's Stationery Office, London. *British History Online*, http://www.british-history.ac.uk/cal-state-papers/venice/vol4/pp278-281.

Bryson, B. 2008. *Shakespeare*. HarperCollins, London.

Bucknill, Sir J. C. 1860. *The Medical Knowledge of Shakespeare*. Longman & Co., London.

Bucknill, Sir J. C. 1867. *The Psychology of Shakespeare*. Macmillan & Co., London.

Bynum, W. 2000. Phrenitis: What's in a Name? *The Lancet*, 356: 1936.

Cassidy, C., Doherty, P. 2017. *And Then You're Dead: A Scientific Exploration of the World's Most Interesting Ways to Die*. Allen & Unwin, London.

Cerasano, S. P. 1998. Edward Alleyn's 'Retirement' 1597–1600. *Medieval and Renaissance Drama in England*, 10: 98–112.

Chaucer, G., ed. Richard Morris. 1866. *The Poetical Works of Geoffrey Chaucer: Volume 5*. Bell & Daldy, London.

Crisp, A. H. 1996. The Sleepwalking/Night Terrors Syndrome in Adults. *Postgraduate Medical Journal*, 72: 599–604.

Davis, F. M. 2000. Shakespeare's Medical Knowledge: How Did He Acquire It? *The Oxfordian*, 3: 45–58.

Dickson, A., Staines, J. 2016. *The Globe Guide to Shakespeare*. Profile Books, London.

Dillon, J. 2006. *The Cambridge Introduction to Early English Theatre*. Cambridge University Press, Cambridge.

Dimsdale, J. E. 2009. Sleep in *Othello*. *Journal of Clinical Sleep Medicine*, 5: 280–281.

Dobson, M. J. 1989. History of Malaria in England. *Journal of the Royal Society of Medicine*, 82(17): 3–7.

Dreisbach, R. H., Robertson, W. O. 1987. *Handbook of Poisoning: Twelfth Edition*. Appleton & Lange, Connecticut.

Duncan-Jones, K. 2004. *Shakespeare's Life and World*. The Folio Society, London.

Dyce, A. (ed.). 1829. *The Works of George Peele Collected and Edited with Some Account of His Life and Writings, Volume 2*. William Pickering, London.

Elsom, D. M. 2015. *Lightning: Nature and Culture*. Reaktion Books, London.

Emsley, J. 2006. *Vanity, Vitality, and Virility: The Science Behind the Products You Love to Buy*. Oxford University Press, Oxford.

Everson, C. A., Bergmann, B. M., Rechtschaffen, A. 1989. Sleep Deprivation in the Rat: III. Total Sleep Deprivation. *Sleep*, 12(1):13–21.

Fabricius, J. 1994. *Syphilis in Shakespeare's England*. Jessica Kingsley Publishers, London and Bristol, Pennsylvania.

Falk, D. 2014. *The Science of Shakespeare: A New Look at the Playwright's Universe*. St Martin's Press, New York.

Furman, Y., Wolf, S. M., Rosenfeld, D. S. 1997. Shakespeare and Sleep Disorders. *Historical Neurology*, 49: 1171–1172.

Gianni, M., Dentali, F., Grandi, A. M., Sumner, G., Hiralal, R., Lonn, E. 2006. Apical Ballooning Syndrome or Takotsubo Cardiomyopathy: A Systematic Review. *European Heart Journal*, 27: 1523–1529.

Girard, R. 1990. Sacrifice in Shakespeare's *Julius Caesar*. *Salmagundi*, 88/89: 399–419.

Greenblatt, S. 2005. *Will in the World: How Shakespeare Became Shakespeare*. Pimlico, London.

Greene, R. 1870. *Greene's Groats-Worth of Wit, Bought with a Million of Repentance: Reprinted from an Original Copy of the Extremely Rare Edition of 1596, Preserved in the Library of Henry Huth, Esq.* Chiswick Press, London.

Gregory, A. 2018. *Nodding Off: The Science of Sleep from Cradle to Grave*. Bloomsbury, London.

Griffiths, R. A. 1969. The Trial of Eleanor Cobham: an Episode in the Fall of Duke Humphrey of Gloucester. *Bulletin of The John Rylands Library*, 51(2): 381–199.

Gurr, A. 2012. *The Shakespearean Stage 1574–1642*. Cambridge University Press, Cambridge.

Hancock, P. A. 2011. *Richard III and the Murder in the Tower*. History Press, Gloucester.

Harries, M. 2003. Near Drowning. *BMJ*, 327: 1336–1338.

Hayden, D. 2003. *Pox: Genius, Madness, and the Mysteries of Syphilis*. Basic Books, New York.

Haynes, A. 1999. *Sex in Elizabethan England*. Wrens Park Publishing.

Heaton, K. W. 2006. Faints, Fits, and Fatalities from Emotion in Shakespeare's Characters: Survey of the Canon. *BMJ*, 333: 1335–1338.

Heyman, P., Simons, L., Cochez, C. 2014. Were the English Sweating Sickness and the Picardy Sweat Caused by Hantaviruses? *Viruses*, 6: 151–171.

Heywood, T. 1579. *An Apology for Actors*. Thomas Woodcocke, London.

Hirschfeld, H. 2018. *The Oxford Handbook of Shakespearean Comedy*. Oxford University Press, Oxford.

Holinshed, R. 1808 edn. *Chronicles of England, Scotland and Ireland*. Johnson, London.

Homer. 1995 edn. *The Iliad*. Wordsworth Editions Limited, Hertfordshire.

Hurren, E. T. 2016. *Dissecting the Criminal Corpse: Staging Post-Execution Punishment in Early Modern England*. Springer, London.

Hutton, A. 1892. *Old Sword Play: The Systems of Fence in Vogue During the XVIth, XVIIth, and XVIIIth Centuries with Lessons Arranged from the Works of Various Masters*. H. Grevel & Co., London.

Jenner, R., Undheim, E. 2017. *Venom: The Secrets of Nature's Deadliest Weapon*. Natural History Museum, London.

Jolliffe, D. M. 1993. A History of the Use of Arsenicals in Man. *Journal of the Royal Society of Medicine*, 86: 287–289.

Jonson, B. 2008. *The Alchemist and Other Plays*. Oxford University Press, Oxford.

Karim-Cooper, F., Stern, T. (eds). 2016. *Shakespeare's Theatres and the Effects of Performance*. Bloomsbury, London.

Kaufmann, M. 2017. *Black Tudors: The Untold Story*. Oneworld Publications, London.

Kelly, J. 2013. *The Great Mortality: An Intimate History of the Black Death*. Harper Perennial, London.

Kerwin, W. 2005. *Beyond the Body: The Boundaries of Medicine and English Renaissance Drama*. University of Massachusetts Press, Massachusetts.

Klaassen, C. D. (ed.). 2013. *Casarett & Doull's Toxicology: The Basic Science of Poisons*. McGraw-Hill Education, New York, Chicago, San Francisco.

Knell, R. J. 2003. Syphilis in Renaissance Europe: Rapid Evolution of an Introduced Sexually Transmitted Disease? *Proceedings of the Royal Society of London, B*, 271: S174-S176.

Kyd, T. 1615. *The Spanish Tragedy, Or Hieronimo is Mad Again*. Mr Dodley, London.

MacDonald, M. 1989. The Medicalization of Suicide in England: Laymen, Physicians, and Cultural Change, 1500–1870. *Milbank Quarterly*, 67(1): 69–91.

MacGregor, N. 2014. *Shakespeare's Restless World: An Unexpected History in Twenty Objects*. Penguin Books, London.

McNeill, W. H. 1976. *Plagues and Peoples*. Anchor Press/Doubleday, New York.

Magnusson, M. 2000. *Scotland: A History of a Nation*. HarperCollins, London.

Marlowe, C. 2008. *Doctor Faustus and Other Plays*. Oxford University Press, Oxford.

Mendilow, A. A. 1958. Falstaff's Death of a Sweat. *Shakespeare Quarterly*, 9(4): 479–483.

Mortimer, I. 2009. *1415: Henry V's Year of Glory*. Vintage Random House, London.

Nash, T. 1596. *Have With You To Saffron Walden: or, Gabriell Harvey's Hunt Is Up*. John Danter, London.

Nashe, T. 1815 edn. *Christ's Tears Over Jerusalem*. Longman, Hurst, Rees, Orme, & Brown, London.

Neely, C. T. 1991. 'Document in Madness': Reading Madness and Gender in Shakespeare's Tragedies and Early Modern Culture. *Shakespeare Quarterly*, 42(3): 315–338.

Neill, M., Schalkwyk, D. (eds). 2018. *The Oxford Handbook of Shakespearean Tragedy*. Oxford University Press, Oxford.

Nicoll, A. (ed.). 1964. *Shakespeare in His Own Age: Shakespeare Survey 17*. Cambridge University Press, Cambridge.

Norwich, J. J. 1999. *Shakespeare's Kings*. Faber & Faber Ltd, London.

Nuland, S. B. 1995. *How We Die: Reflections on Life's Final Chapter*. Vintage Books, New York.

Öğütcü, M. 2016. Public Execution and Justice On/Off the Elizabethan Stage: Shakespeare's First Tetralogy. *Mediterranean Journal of Humanities*, VI/2: 361–379.

Oliver, N. 2011. *A History of Scotland*. Orion Books Ltd, London.

Orent, W. 2004. *Plague: The Mysterious Past and Terrifying Future of the World's Most Dangerous Disease*. Free Press, New York.

Ovid, trans. Riley, H. T. 1858. *The Metamorphoses of Ovid*. H. G. Bohn, London.

Paster, G. K. 1993. *The Body Embarrassed: Drama and Disciplines of Shame in Early Modern England*. Cornell University Press, New York.

Pauli, R. (ed). 1857. *Confessio Amantis of John Gower*. Bell & Daldy, London.

Pliny, trans. Bostock, J., Riley, H. T. 1856. *The Natural History of Pliny*. G. Bell & Sons, London.

Plutarch, trans. Langhorne, J., Langhorne, W. 1850. *Plutarch's Lives, Translated from the Original Greek: With Notes, Critical and Historical and the Life of Plutarch*. Applegate Publishers, London.

Pollard, A. J. 1995. *Richard III and the Princes in the Tower*. Alan Sutton Publishing Ltd, Gloucestershire.

Prahlow, J. 2010. *Forensic Pathology for Police, Death Investigators, Attorneys, and Forensic Scientists*. Springer, London.

Prestwich, M. 1996. *Armies and Warfare in the Middle Ages: The English Experience*. Yale University Press, New Haven and London.

Quigley, C. 2015. *The Corpse: A History*. McFarland & Co., North Carolina.

Reiter, P. 2000. From Shakespeare to Defoe: Malaria in England in the Little Ice Age. *Emerging Infectious Diseases*, 6(1): 1–11.

Retief, F. P., Cilliers, L. 2005. The Death of Cleopatra. *Acta Theologica Supplementum* 7, 26(2): 79–88.

Roach, M. 2003. *Stiff: The Curious Lives of Human Cadavers*. Penguin Books, London.

Ross, J. J. 1965. Neurological Findings After Prolonged Sleep Deprivation. *JAMA Neurology, Archives of Neurology*, 12: 399–403.

Sakai, A. 1991. Phrenitis: Inflammation of the Mind and Body. *History of Psychiatry*, 2: 193–205.

Saukko, P., Knight, B. 2004. *Knight's Forensic Pathology: 3rd Edition*. Hodder Arnold Ltd, London.

Sawday, J. 1996. *The Body Emblazoned: Dissection and the Human Body in Renaissance Culture*. Routledge. London.

Schoenbaum, S. 1987. *William Shakespeare: A Compact Documentary Life*. Oxford University Press, Oxford.

Scott-Warren, J. 2003. When Theatres Were Bear-Gardens: Or, What's at Stake in the Comedy of Humours. *Shakespeare Quarterly*, 54(1): 63–82.

Shakespeare, W., ed. Bate, J., Rasmussen, E. 2008. *William Shakespeare: Complete Works*. Random House, London.

Sharpe, J. A. 1999. *Crime in Early Modern England 1550–1750*. Longman Ltd, London, New York.

Simrock, M. K. 1850. *The Remarks of M. Karl Simrock, on the Plots of Shakespeare's Plays: With Notes and Additions by J. O. Halliwell*. The Shakespeare Society, London.

Smith, A. 2020. 'See How the Blood Settled in His Face': Shakespeare's Warwick – Fiction's First Pathologist. *Brief Encounters*, 4(1): 82–90.

Smith, M. 1992. The Theatre and the Scaffold: Death as Spectacle in *The Spanish Tragedy*. *Studies in English Literature, 1500–1900*, 32(2): 217–232.

Steinmetz, A. 1868. *The Romance of Duelling in All Times and Countries: Volume 1*. Chapman & Hall, London.

Stubbes, W. 1836 edn. *The Anatomie of Abuses*. W. Pickering, London.

Tadros, G., Jolley, D. 2001. The Stigma of Suicide. *The British Journal of Psychiatry*, 179(2): 178.

Taylor, J. E. 1855. *The Moor of Venice: Cinthio's Tale and Shakespeare's Tragedy*. Chapman & Hall, London.

Teske, A. J., Verjans, J. W. 2016. Takotsubo Cardiomyopathy – Stunning Views on the Broken Heart. *Netherlands Heart Journal*, 24: 508–510.

Thomas P. G. (ed.). 1907. *Greene's 'Pandosto' or 'Dorastus and Fawnia' Being the Original of Shakespeare's 'Winter's Tale'*. Chatto & Windus, London.

Thompson, C. J. S. 1993. *Poisons and Poisoners*. Barnes & Noble Books, New York.

Valentine, C. 2017. *Past Mortems: Life and Death Behind Mortuary Doors*. Little, Brown, London.

Verbruggen, J. F. 2002. *The Art of Warfare in Western Europe During the Middle Ages*. Boydell Press, Suffolk.

Webster, J. 2009. *The Duchess of Malfi and Other Plays*. Oxford University Press, Oxford.

Weir, A. 2009. *Lancaster and York: The Wars of the Roses*. Vintage Random House, London.

Wells, S. 2006. *Shakespeare and Co*. Penguin Books, London.

Williams, N. 1957. *Powder and Paint: A History of the Englishwoman's Toilet Elizabeth I–Elizabeth II*. Longmans, Green & Co., London.

Wilson, D. 2014. *The Plantagenets: The Kings that Made Britain*. Quercus Editions, London.

Websites

british-history.ac.uk/letters-papers-hen8
opensourceshakespeare.org
plato.stanford.edu/entries/death-definition

TV

The Shakespeare Collection. BBC DVD.

Acknowledgements

So, thanks to all at once and to each one

Macbeth, Act 5, Scene 8

First of all thanks to Jim Martin, Anna MacDiarmid and everyone at Bloomsbury for another opportunity to write about something I love and for all their support throughout. Thanks also to Catherine Best for her brilliant editorial skills. Ele Willoughby also deserves a huge amount of praise for her superb chapter illustrations.

I wrote this book out of an enthusiasm for Shakespeare and the gorier side of science. My background in chemistry meant the research into scientific aspects were much less daunting than the literary and theatrical side of things. Matt Pinches of the *Guildford Shakespeare Company* and Katherine Mendelsohn have been brilliant in helping me understand more about how Shakespeare is performed – thank you.

On the science side of things, Margaret Skinner and Caroline Barrett helped me out of a difficult spot regarding animal blood. Alice Gregory supplied some great insights into sleep, or the effects of a lack of it. Isabelle Sheridan has been fantastic at answering my questions about anatomical and medical details. Bill Backhouse and Justin Brower need a special mention for helping track down sources and resources for me.

Thanks to David and Sharon Harkup, Beatriz Gonzáles, Matt May, Helen and Andrew Skinner, Richard Stutely, Mark Whiting and all in the Valencia Writing Group, particularly Dónal Mac Erlaine, and most of all to my parents, for reading and commenting on the book as it progressed. The result has been much improved by their feedback. Thanks also to Burton King, Andrea Smith and Rob Shennan for pointing out some errors in the hardback edition and possible improvements for the paperback. Though many people have helped spot and correct my mistakes nothing is perfect. Any remaining errors are mine and mine alone.

Obviously this book owes everything to William Shakespeare and his splendid work. 'And how quote you my folly?' Most quotes, along with a lot of additional information, have been taken from the excellent *Open Source Shakespeare* website (www.opensourceshakespeare.org). All quotes from *The Two Noble Kinsmen* have come from the RSC edition of the *Complete Works of William Shakespeare*. And finally, quotes from *Edward III* were found in John Julius Norwich's *Shakespeare's Kings*.

Index

accidental injuries 32, 52, 54
aconitine 249, 267
actors 16, 24, 33, 34, 38, 54, 57
  acting styles 57–58
  swordfights 172–73, 190
Agincourt, battle of 174, 177, 180, 183–89, 191, 195, 196–201, 203, 221
air embolism 193
alcoholism 22, 41, 208
All's Well That Ends Well 77, 80, 210, 293, 337, 342
Antony and Cleopatra 57, 60, 95, 134, 237–38, 254–60, 271, 273, 285, 298–99, 350
apothecaries 74, 75, 250
arrow injuries 186–89, 191–92, 193, 267
arsenic 239, 242, 243, 254
artistic licence 59–60
As You Like It 9, 11, 28, 337, 342
asphyxia 123, 144, 145, 149, 290
atropine 249, 263
attitudes to death 10, 73, 338
audiences 48, 49, 52, 55, 57, 61

The Battle of Alcazar 29, 62, 66
bear attacks 333–36
bear-baiting 45, 49, 52, 61, 330–33
Beaumont, Francis 40
beheading 45, 67, 115, 117–18, 120, 124–28, 142, 163
benefit of clergy 34, 110
benefit of the womb 112
bleeding to death 16, 125, 194, 195, 216, 273, 274, 275, 323
blood
  circulation of 75
  congealed 158, 276
  pooling 65, 101, 133, 143, 202
  as a theatre prop 60, 64–70
body parts as theatre props 60–64
boiling 240
Bosworth, battle of 168–69
brain 84–85, 90, 91, 96, 101, 125–26, 150, 151, 192, 196, 295, 309, 310
breathing 84, 88, 89, 90–91, 95–96, 97, 101, 136, 146, 150, 193, 194, 249, 299, 306, 317
burials 18, 42, 82, 92, 200, 288, 289, 292
burning
  by lightning strikes 315, 316, 318–19
  as punishment 122, 124

cadaveric spasms 199
Caesarean section 321
cancer 159
cannabalism 325–28
carbon monoxide poisoning 286
Cardenio 39, 40
carphologia 98
cause of death 101
  determination 82–83, 142–47, 153, 219
  records 82–83

ceruse 238
Chapman, George 34–35
child mortality 13
childbirth 13, 86, 87, 89, 303, 321
cirrhosis of the liver 22
The Comedy of Errors 71, 103, 104, 293, 341
Coriolanus 171, 173, 191, 202, 206, 320, 347
costumes 58–59
crimes
  punishments 103–37
  pardons for 109, 114
crushing 135–37, 189
curare 267
cyanide 251–52, 267
cyanosis 143, 147
Cymbeline 38, 92, 252–54, 297–98, 351

Davenant, Sir William 231, 232
death masks 100
death process 84, 98
death rattle 102, 202
death sentence 34, 104–10
decay 86–87
definitions of death 83–84
dehydration 134
Dekker, Thomas 35, 39, 40
delirium 99, 208, 303, 304
diarrhoea 221, 247
dismemberment 61, 321
doctors, physicians and surgeons 42, 74, 75, 77, 79, 81, 85, 86, 194, 201, 320
drawing and quartering 31, 119, 120, 121
Drayton, Michael 41
dropsy 22
drowning 161, 187, 269, 286–92
The Duchess of Malfi 62, 63
duelling 34, 54, 173
dysentery 115, 182, 221–22, 223, 250

ebony 235, 262, 263
Edward III, King 181, 188, 217
Edward III 10, 175, 243, 351
Edward IV, King 160–61, 162
Edward V, King 162–68
Edward of Westminster, Prince 158, 159
Elizabeth I, Queen 12, 15, 30, 36, 129, 154, 185, 238
ergot 245
erysipelas 100
executions 45, 61, 71, 103–4, 116–37
eyeballs, fake 63–64
eyes 91–92, 101, 279, 283, 297

facial injuries 182, 231
fainting 295–96
fake death 92–97
falls 280–82
fencing 61, 172, 266
fire, theatres destroyed by 51
fleas 71, 212, 213, 214–15, 216
flesh, pound of 320
Fletcher, John 39, 40, 347
food poisoning 19
food shortages 31, 38–39
frenzy 303
friars and monks 93, 94, 164, 220, 223, 246–48

gangrene 100
gibbeting 133–34

Globe Theatre 38, 44, 46, 47, 51–52, 53, 60, 228, 330
gonorrhoea 226
Greene, Robert 20–22, 24, 26, 29
grief 71, 89, 293–312
guaiac wood 233, 235, 264
Gunpowder Plot 126, 312

haemorrhage 13, 143, 196, 281
Hall, John 42, 76
hallucinations 248, 249, 305
Hamlet, Prince of Denmark 15, 43, 48, 49, 50, 58, 60, 61, 106, 235, 240–41, 261–70, 286–92, 301, 339, 349
hanging 22, 31, 45, 106, 118–21, 128–33, 152, 153
hanging in chains *see* gibbeting
hantavirus 207
Harfleur, siege of 181, 182, 222
Harvey, William 75, 76
head, blow to the 150–51, 195–96
heads, decapitated 45, 46, 127–28
heart and cardiac deaths 84, 88, 276, 296, 297, 299–300, 317
hellebore 262
hemlock 244, 262
henbane 262–63
Henry IV, King 99–100, 134, 179
Henry IV 99, 172
  Part I 119–20, 129, 176, 179, 192, 215, 228, 242, 343, 353
  Part II 76, 78, 95, 98, 205, 227, 242, 249, 295, 344
Henry V, King 127, 203, 222, 223; *see also* Agincourt, battle of
Henry V 32, 46, 56–57, 61, 72, 98, 127, 129, 172, 174, 179–80, 184, 188, 191, 195, 196–201, 209, 221–23, 234, 344
Henry VI, King 156–57, 158, 203
Henry VI 55, 154, 172, 174, 185
  Part I 28, 112, 122, 172, 181, 185, 233, 306, 345
  Part II 62, 95, 100, 102, 110, 119, 120–22, 125, 141–43, 155, 178, 246, 304, 306, 345–46
  Part III 140, 148, 155, 178, 193, 346
Henry VII, King 169
Henry VIII, King 190, 228
Henry VIII, Or All Is True 39, 51, 77, 101, 117, 300–1, 347
Holinshed's *Chronicles* 111, 135, 147, 152, 159, 160, 163, 177, 249–50, 283, 303, 311
hypercarbia 146
hypothermia 87, 88

imprisonment 33, 34, 40, 114–15, 121
infanticide 122
influenza 41, 206, 207

James I, King 37–38, 82, 311–12
The Jew of Malta 28, 265
John, King 223, 247–50
Jonson, Ben 21, 24, 32–5, 41, 57, 330
Julius Caesar 49, 68–69, 70, 75, 272–73, 276, 284–86, 328–29, 348–49

King John 175–76, 223, 247–50, 278–83, 300, 301–4, 343
King Lear 71, 90–91, 151–53, 239, 241–42, 295, 297, 333, 350
King's Men 37, 68
Kyd, Thomas 27, 28, 106, 217, 292

legal processes 113–14
leprosy 100, 206, 207, 264, 303
Lichtenburg figures 318
life expectancy 12, 17
lightning 314–20
London
  crowding and insanitary conditions 17–20
  Great Stone Gate 45, 46, 127, 128
  plague epidemics 23–24, 217–19
  see also theatres
Lord Chamberlain's Men 25, 30, 36, 37
A Lover's Complaint 71, 352
Love's Labour's Lost 71–72, 214, 295, 341
lye 67, 248

Macbeth, King of Scots 311–12
Macbeth 54, 60, 81, 122, 133, 191, 217, 242–45, 269, 295, 300, 301, 305–8, 310–12, 321, 332, 339, 349
madness 287, 288, 302, 303
malaria 99, 208–9
malnutrition 31, 115
mandragora 95
Marlowe, Christopher 20, 24, 26–8, 29, 35, 265, 345
Marston, John 35
Measure for Measure 39, 71, 104–5, 113, 115, 207, 235, 341
medical treatments 79–81, 85, 194–95, 201–2, 220–21, 322
medicines 88, 92, 94–96, 209, 234–35, 239
The Merchant of Venice 28, 108, 237, 320, 342
mercury 81, 234–35, 238, 264
mercy 108–9
The Merry Wives of Windsor 30, 71, 76, 77, 98, 172, 215, 333, 341
Middleton, Thomas 35, 39, 40, 341, 348, 349
A Midsummer Night's Dream 277, 340, 13, 340, 341–42
midwives and wise women 12–13, 73–74, 75, 77, 247
mob violence 328–29
mortality statistics 13, 23, 82
Much Ado about Nothing 207, 295–96, 341
murder 26, 28, 30, 34, 67, 139–69, 239, 240, 242, 254, 266, 271, 287, 295, 305, 312, 322, 328

Nashe, Thomas 20, 29, 33, 332, 345

Othello, the Moor of Venice 95, 148–50, 194, 271, 278, 298, 350

pain relief 80, 321
pain threshold 90, 192
pallor 89, 143, 295
palsy 147
Peasants' Revolt 111

Peele, George 29, 62, 66
Pericles, Prince of Tyre 39, 48, 57, 86, 295, 314–20, 351
perimortem 153
peritonitis 274
personal hygiene 20, 49–50, 215
petechiae 144
The Phoenix and the Turtle 352
plague 12, 16, 22, 24, 37, 40, 82, 210–21; *see also* theatre closures
play length, time restrictions on 56
pneumothorax 189
poems 24, 71, 274–76, 233, 234, 352
poisons and poisoning 25, 75, 80, 93, 96–97, 159, 235, 237–68, 305
population 12, 17
pox *see* syphilis
pressing *see* crushing
Princes in the Tower *see* Edward V, King *and* Richard of Shrewsbury
props 60–70, 125, 319, 334
prostitutes 45, 72, 227–28, 234, 235
pulse 79, 91, 94, 146, 202, 252

rape 274–75, 322
The Rape of Lucrece 274–76, 352
revenge 106, 114, 140, 218, 266, 286, 321, 323, 325, 328, 333
revival from apparent death 85–90; *see also* fake death
Richard II, King 111, 134, 135, 188
Richard II 30, 134, 137, 173, 244, 294, 343
Richard III, King 153–69
Richard III 153–69, 174, 346
Richard of Shrewsbury 162–68
rigor mortis 90, 145, 199
risus sardonicus 304–5
Romeo and Juliet 28, 30, 56, 92–5, 109, 172–73, 195, 211, 220, 246–47, 250–52, 271, 297, 300, 348

scarlet fever 206, 207
scoliosis 154
scrofula 224
self-harming 284
sepsis 202–3
septic shock 203, 213
Shakespeare, William
  birth and early life 11–12, 14
  coat of arms 36, 37
  companies and theatres 25, 30, 36–37, 44, 51–52
  education 21
  family, marriage and children 11, 14, 18, 36, 76, 301
  leaving Stratford for London 15–17
  return to Stratford, death and burial 40–42
Shakespeare's works
  collaborations 28, 29, 39–40
  disputed authorship 76
  First Folio 57
shock 194
signs of death 90–92
Sir Thomas More 39
skin changes 101–2
sleep conditions 305–10
smallpox 209, 214, 224
smells, foul 49–50, 318–19
snakebite 256–60, 265–66

soldiers 32, 177, 179–80, 223, 227
Sonnets 233, 234, 352
The Spanish Tragedy 106
stabbings 16, 27, 65–6, 93, 157, 272–78
starvation 31, 134, 135, 219
strangulation 143–5, 148, 153
street fights 16, 26, 29, 173
stress 192, 299–300, 306
strokes 99, 100, 147
strychnine 305
suffocation 148, 153, 190
suicide 256, 257, 269–92
swallowing fire 285–86
sweating sickness 98, 99, 206–7
swelling 237–38, 247, 266
sword fighting 16, 26, 30, 54, 172–73, 190, 266
sword injuries 190–91, 193
syphilis 22, 29, 41, 72, 81, 206, 209, 223–36, 264

The Taming of the Shrew 302, 308, 342
The Tempest 48, 205, 208, 210, 341
tetanus 305
tetrodotoxin (TTX) 96–97
theatre closures 23, 31, 38, 44, 48, 211, 219
theatre companies 15–16, 23, 24–25, 38, 52–53, 55, 69
theatres 43–70
thrombosis 182
Timon of Athens 39, 80, 218, 348
Titus Andronicus 29, 57, 59, 62, 135, 139, 300, 321–28, 329, 347–48
tourniquets 194, 322
Troilus and Cressida 214, 228–29, 232, 236, 347
tuberculosis 100, 159, 206, 207
Twelfth Night, Or What You Will 36, 48, 50, 71, 114, 172, 210, 213, 238, 244, 293, 313, 332, 333, 342
Two Gentlemen of Verona 78, 341
The Two Noble Kinsmen 39–40, 114, 351
typhoid 19, 41–42, 99, 206

uremia 202
urination, painful 233, 265
urine 50, 67
urinoscopy 78

vagal reflex 133
venom 240, 242, 243, 255–60, 265–66
Venus and Adonis 24, 352
vomiting 79, 212, 233, 239, 247, 249, 252

war 171–203
Watson, Thomas 26
weather 48–49
Webster, John 35, 62, 63
Wilkins, George 39
The Winter's Tale 63, 89–90, 112, 295, 296–97, 333–35, 342
witches and witchcraft 25, 122, 136, 242–45, 312
wounds
  infected 201–3
  re-opening 158
  sucking 189, 274

yaws 226
yew 244